IB BIOLOGY

Internal Assessment

IB BIOLOGY

Internal Assessment

For the International Baccalaureate Diploma

Zouev Elite Publishing

Published 2021

Printed by Zouev Elite Publishing

ISBN 978-1-9996115-3-8, paperback.

TABLE OF CONTENTS

PART I
THE BIOLOGY IA GUIDE

1. GENERAL INTRODUCTION

The Internal Assessment in Biology is a personal research that comprises the 21st century skills in Science; research skills, thinking and communication skills, analytical skills as well as self-management and time-management. As a component, it weighs 20% of the total grade, it is externally graded but it is, initially, moderated internally by the teacher.

Biology is a Life Science, which allows you to be a risk taker and inquisitive. Hence, you can carry out any type of investigation, even beyond the syllabus, given does not challenge ethical or environmental issues.

Tip: Going beyond the syllabus shows serendipity and courage but you must be careful not to choose a topic that demands skills you have not acquired or a method that is too complex to be carried out in a school setting. IB allows students to carry out an investigation in an external lab such as university lab, as long as there is validation and proof.

1.1 The Criteria of Internal Assessment

There are 5 criteria that are tabulated below. The approach in overall marking is the best-fit approach, which means that the moderators will mark positively to what you have achieved overall. The total number of marks is 24.

Criterion and marks	What is marked?
Personal Engagement (PE) – 2 marks	Your curiosity, your input and insight in the topic, your creativity or serendipity. It is marked by the whole investigation, not just the introduction
Exploration (Ex) – 6 marks	The background theory, the research question, the method, the variables and their manipulation.
Analysis (A) – 6 marks	The presentation of raw data, the manipulation of numbers and significant figures, dealing with uncertainty, graphical

	analysis, data processing, statistical processing, interpretation of data.
Evaluation (Ev) – 6 marks	This criterion comprises the Conclusion, addressing strengths and limitations, suggesting viable improvements and future extensions.
Communication (C) – 4 marks	The general flow and coherence of your report, the clarity and presentation of the graphs and tables, the use of scientific terminology, proper referencing etc.

1.2 The layout of the IA report

The IA report should be written in either 11 or 12 font size. Make sure you keep the spacing between 1.15 and 2.0. Avoid making it colorful as this is not graded, and it may be not likeable by the moderator. The report should not exceed the 12 pages, although 13 pages may be acceptable if page 13 contains part of the bibliography. Generally, avoid, too densely written text. As for the Appendix, this part is never checked or graded by the moderator, so it is not a good idea to put important data there. However, if you have lengthy raw data tables, you may keep some sample tables in the main body and add the rest in an appendix; in this case the teacher must point out in the comment of your report, prior to submission, that they have checked your appendix.

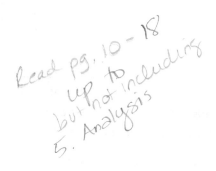

Read pg. 10 - 18
up to
but not including
5. Analysis

2. FINDING A GREAT TOPIC

This is probably the most exciting part of your investigation. However, it can be overwhelming and lead to either a very complex or a very simplistic research question. So, here are the steps you are advised to take to find a topic that excites you and will lead to a meaningful research question:

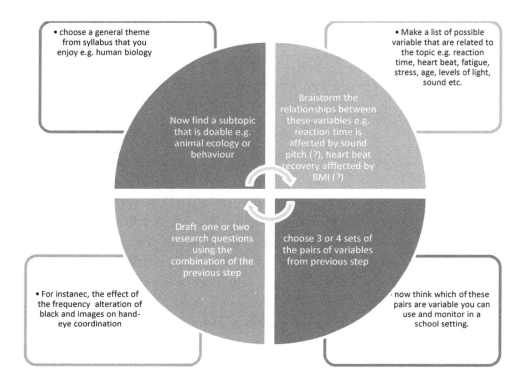

• choose a general theme from syllabus that you enjoy e.g. human biology

Now find a subtopic that is doable e.g. animal ecology or behaviour

• Make a list of possible variable that are related to the topic e.g. reaction time, heart beat, fatigue, stress, age, levels of light, sound etc.

Braistorm the relationships between these variables e.g. reaction time is affected by sound pitch (?), heart beat recovery afffected by BMI (?)

Draft one or two research questions using the combination of the previous step

choose 3 or 4 sets of the pairs of variables from previous step

• For instanec, the effect of the frequency alteration of black and images on hand-eye coordination

now think which of these pairs are variable you can use and monitor in a school setting.

TIP: If your investigation involves human subjects, you must obtain a consent form. If your investigation involves animals (including arthropods), then make sure your method abides by the protocol of ethical treatment of animals as laid out by IB.

2.1 Choosing the Source Of Raw Data

A vital factor that may affect the choice of topic and research question is how you are planning to collect raw data. There are two types of data collection: primary data collection and secondary data collection.

i.   Primary data collection

Data is collected with an actual, hands-on experiment in the lab or a field, or from a simulation. Here are great resources:

https://www.wolfram.com/system-modeler/libraries/high-school-biology/?src=google&458&gclid=Cj0KCQiA1KiBBhCcARIsAPWqoSqVc-pvfE7oElp_v7N0G_PXGp2ypiVh6IHPEdOJQKxCdtSNYlqrKlQaAvVzEALw_wcB

https://phet.colorado.edu/en/simulations/filter?subjects=biology&type=html

https://i-biology.net/ict-in-ib-biology/modeling-simulation/

https://www.sciencecourseware.org/FlyLabJS/

ii.   Secondary data collection

Data is collected from a data base. It is advisable you receive feedback from your teacher as to which data base will provide sufficient data for a focused research question:

DNA Data Bank of Japan

UniGene

Explore worm Biology

Genomes, Genome features and maps

Protein Data Bank

http://www.rcsb.org

3. PERSONAL ENGAGEMENT

The criterion of PE applies holistically to the whole body of the lab report. The areas that are assessed are seen below:

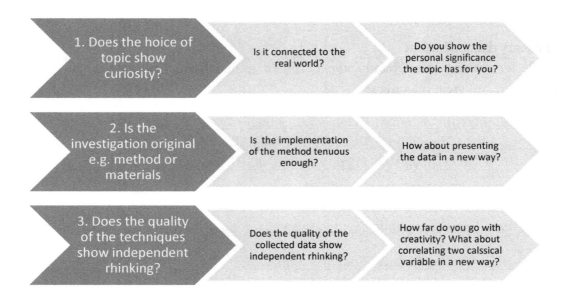

Tip: It is advisable that you write a clear statement of the purpose of the investigation. For instance:

The concerning accumulation of painkillers in waste water, such as ibuprofenoic acid, inspired me to investigate the effect of this drug on the growth of Phaseolus vulgaris.

4. EXPLORATION

The criterion of exploration consists of the following components: the background theory, the research question, the variables, the risk assessment and acknowledgment of limitations.

4.1 The background theory

The background information provided can enhance significantly your report as long as it is focused on the research question, it has plenty of references and it is not too general or simplistic. Normally, a well written background theory should take up 1.5 to 2 pages.

Do's	Don'ts
The information is pertinent to the aim and the research question. For instance, if your study is about an enzyme, gather information on the specific enzyme and the factors that may affect it, especially the factor used as IV.	To include encyclopedic and general information. Mentioning the general background of how enzymes function even if this may not be relevant to the research question, affects, not only the mark for Exploration, but also for Communication.
Use scientific information from scientific articles and journals. Try to go as deep as possible without making the topic too complex. Use in-text citations	Sources such as popular magazines and Wikipedia, are not considered appropriate sources of information. Avoid footnotes, except to clarify a new or unfamiliar term etc.

Tip: You may use picture or figures, with a reference, but make sure they are absolutely relevant to the research question. E.g. the picture of an the specific enzyme's shape is appropriate but the general figure of an enzyme's active site is not a good idea.

4.2 Developing the Research Question

A **focused** research question (RQ) must have these ingredients:

- It must be presented early in the lab report.
- It could be ideally, of the format "To what extent the change in X affects the Y as measured by....".
- The independent variable must be clearly stated with proper units and the full range and intervals.
- The dependent variable must also be stated clearly with units
- A reference to the method of measurement
- If there is an organism involved, the scientific name of the organism in italics should be stated: e.g. instead of beans you must write *Phaseolus vulgaris.*
- Do not compare organisms or products.

Here are some examples of research questions with comments from the moderator:

Research Question	Moderator's comments
How the pH affects the growth of Lentils plant.	*The RQ is not focused. It is not clear whether it is the pH of the water or the soil, also what is meant by 'growth'.*
To what extent the pH of the soil affects the height of lentils plants.	*This RQ is more focused but the units and the scientific name are missing*
To what extent does the pH of the soil ($\pm$0.1: 3.0, 4.0, 5.0, 6.0, 7.0, 8.0, 9.0) affects the rate of growth of *Lens culinaris* as measured by the %change of the height of its stem daily for two weeks.	*This RQ is focused, with clear IV and DV, range and intervals for IV and reference of the measurement.*

4.3. Identifying the Variables

i. The Independent Variable (IV)

It is advisable that the IV is quantifiable, with clear range and intervals stated. Also, the range and the intervals of the IV must be **justified.** Ensure, that the changes in the values of IV are well monitored in the method. Also, any limitations or uncertainties must also be clearly stated.

The intervals of the IV should be such that the data collected is valid. For instance, a change of pH of 1.0 unit (e.g. from 2.0 to 3.0) is more valid than a change of 2.0 units (e.g. from 2.0 to 4.0).

Tip: Often the IV is not quantifiable, which is well accepted, as long as the investigation leads to rigorous data collection. For instance, a student may wish to use a t-test to compare the size of the population of the same species in different habitats.

16

ii. The Dependent Variable (DV)

The dependent variable is a protagonist in the lab report. You should present how it is measured with units and method. Also, you should show how the DV is related to the IV.

Below there are examples of IV and DV with moderator's comments:

The Independent and Dependent variables	Moderator's comments
IV: the pH of the water (2.0-9.0). **DV:** the height of the lentils plant. I will measure the growth of the plant every day for two weeks	*For the IV: the intervals are not stated, nor the uncertainty. Also, there is no mention on how the changes will take place or why this range of pH was chosen.* *For the DV: the scientific name is not mentioned. It is not clear how the growth is measured or calculated. Also, it is not clear at which developmental stage of the plant the growth is measured.*
IV: The pH of the irrigation water (±0.1: 3.0, 4.0, 5.0, 6.0, 7.0, 8.0, 9.0) used for the plants of lens culinaris, right after germination. The pH is controlled with buffers. The range of pH is 2.0-9.0 as this is the commonest values found in cultivated areas under the effect of acid rain or fertilizers. **DV:** the %change of height (Δh) of *Lens culinaris* seedlings measured daily for two weeks. Initially the height of the stem is measured and then the % change of height (Δh) is calculated as: $\%(\Delta h)$ $$= \frac{height\ in\ Day\ 2 - height\ in\ Day\ 1}{height\ in\ day\ 1} \times 100$$	*For the IV: the range and intervals are clear with uncertainty and they are justified. Also, the method of how the pH is monitored, is mentioned.* *For the DV: The exact method of measuring and calculating the DV is explained with the scientific name and the developmental stage of the plant.*

iii. The Control Variables

The control variables should be at least 4 or 5 factors that may affect the validity of the collected data. In Biology these factors usually include temperature, pH, humidity, light intensity, parasites and forth. The control variables ideally could be presented in a table although this is not required. However, there must be a clear in-depth clarification of how these factors may skew the results and how you are planning to monitor or control them. Here are some examples:

Control variable	Potential impact	Method of control
Temperature (°C ± 0.5)	Inconsistency in the values of temperature that the seedlings are exposed to may affect their growth and, hence, skew the results. This could be the source of random error. Seghal et al. (2017) support that the combination of temperature with humidity has significant effect on lentil crops.	All sample pots that contain the growing lentil seedlings are kept in a well-ventilated room with consistent temperature 25.0°C, controlled with the use of air condition.
The variety and genotype of the seedlings	Undeniably the genetic make of the lentil plants are affecting the degree to which a specific variety may respond to changing factors. Hence, if different varieties are used , the deviation in heir response invalidates the collected data.	The seedlings were grown from seeds bought from the same packet assuming that they will all belong to the same genotype.

4.3 The Method

The procedure should be designed to be **reproducible**, and to provide **sufficient** and **reliable** data.

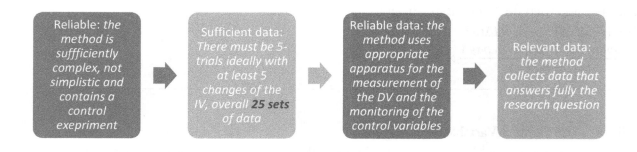

4.4 Risk assessment, environmental and ethical considerations

- Safety issues must be properly addressed as well as the means used to tackle them. For instance, if you used buffers to control the pH, the safety statement could be: *The remaining of the buffer solutions could cause skin and eye irritation so gloves and goggles should be worn while using them. Also, the remaining should not be thrown into the sink; they must be tackled with by the technician according to the national safety protocol.*

- There is a list of specific hazardous microbes that IB does not allow their use for microbial cultures so make sure you get advice from your teacher as to which microbes you can actually use.

- Environmental considerations are of primary importance. For instance, if you do field work, show how you ensured that you had minimum impact on the habitat and that there was no disturbance. If you use animals, consult the IB protocol on the ethical treatment of animals; no creature should be put under stress or pain and definitely animals should be killed. If specimen animals are taken from a habitat, make sure they return back there.

5. ANALYSIS

The Analysis is a quite challenging but fascinating part of your report as it reveals the trends that will help you draw a valid conclusion and answer the Research Question. The components of Analysis include collection, processing and presentation of raw data, propagation of uncertainties, graphical analysis and interpretation of processed data.

Always include a sample calculation for every calculation you carry out!

5.1 Dealing with the Raw Data

The collection of Raw Data means that you collect qualitative and quantitative information. The observations and any kind of quantitative information could be presented prior to the tables with the quantitative raw data collected.

• The Raw Data and Processed Tables

The tables must be numbered with clear headings, units and uncertainties. There should be a title for each table as well. The numbers must have consistency in significant figures with each other but also with the absolute uncertainty as seen in the exemplar table below:

Table 1: Raw data of the time taken for 50 cm³ of oxygen gas to be produced for each trial of yeast, *Saccharomyces cerevisiae*, immobilized on the calcium alginate beads at the different pH values:

pH to which the immobilized yeast was exposed (±0.1 pH)	Time taken for 50 cm³ of oxygen gas to be produced / second (±1 s)				
	Trial 1	Trial 2	Trial 3	Trial 4	Trial 5
2.0	331	347	329	456	332
4.0	107	113	107	109	112
7.0	39	40	41	36	40
9.0	72	98	70	76	71
11.0	396	482	397	404	399

Tip: It is a good idea to centre the numbers in the columns.

5.2 Methods of Data processing

There must be a sample calculation for every processing step you adopt for your study. Here are some of the most important calculations in a Biology IA.

• Calculation of the mean

The mean value of a set of trials can be calculated using the GDC, Excel or Google sheets. The mean value must have the same number of significant figures and decimal places as the individual values. The absolute uncertainty of the mean should be the max-mean range or the standard deviation.

• The max-min range

The max-mean range is an advisable calculation if you have obtained 3 sets of data or anything less than 5. For instance, for 3 values of height, 3.20 cm, 3.80 cm and 4.15 cm, the max-min range is:

$$\frac{\max value - \min value}{2} = \frac{4.15 - 3.20}{2} = \pm 0.48 \text{ cm}$$

- **Uncertainty propagation**

Uncertainties in Biology are important when they highlight the interpretation of how the life science approaches a specific research question. Hence, for a Biology IA the uncertainty propagation does not need to be sophisticated; it should be just enough to show the range of certainty for the conclusions drawn.

$$\% \text{ uncertainty} = \frac{absolute\ uncertainty\ or\ SD}{value} \times 100$$

When two values are subtracted, the absolute uncertainties are added, e.g.:

a= 3.00 $\pm$ 0.05 and b= 2.00$\pm$ 0.05, then a-b = 1.00 $\pm$0.10

If a dependent variable is calculated from two division or multiplication of other values that hold their own uncertainties, then the % uncertainties are added. For instance:

%uncertainty of a= $\frac{0.05}{3.00} \times 100$ = 1.2% and %uncertainty of b=$\frac{0.05}{2.00} \times 100 = 2.5\%$

So, if x=$\frac{a}{b}$ or x= a*b, then %uncertainty of x=%uncertainty of a + %uncertainty of x= 1.2%+2.5%=3.7% or 4%.

- **The standard deviation**

For the calculation of the standard deviation, which can be completed on GDC, Excel or Google sheets, you need a set of 5 values at least. Here is an example:

Time taken for 50 cm^3 of oxygen gas to be produced/seconds ($\pm$1 s)				
Trial 1	Trial 2	Trial 3	Trial 4	Trial 5
453	462	451	448	451

<u>Formula for the calculations of the standard deviation:</u>

$$\sigma = \Sigma \sqrt{\frac{(x-mean)^2}{the\ number\ of\ trials}} = \pm 54.7s, \text{rounded to} \pm 58 \text{ s}.$$

• The change (Δx) and the % change (%Δx)

A common error in processing is that the candidate calculates the change, such increase as decrease of a value, without estimating the change as a %. The calculation of the %change gives a clear trend, a more reliable picture of the change that goes beyond the actual values. Here is an example, for a set of data about the height of seedlings vs. lime content:

%w/w of lime content	Height of seedlings (cm ±0.1)					%increase in height
	Day 1	Day 2	Day 3	Day 4	Day 5	
0.0	1.0	1.0	1.2	1.3	1.3	6
10.0	1.0	1.2	1.2	1.3	1.8	16

Sample calculation for % increase in height at 10.0%w/w in Set 1=

$$((\frac{height\ in\ day\ 5-height\ in\ day\ 1}{height\ in\ day\ 1}) \div 5) \times 100 = ((\frac{1.8-1.0}{1.0}) \div 5) \times 100 = 16\%$$

This is a very convenient calculation as one can graph the %change (y-axis) vs. the IV (x-axis).

• Calculation of the rate

This type of calculation is quite sufficient when it comes to changes observed with time, such as change in length, height, volume of gas produced or consumed, pH etc. Rate is a change in a variable (DV) per unit of time:

$$\text{Rate} = \frac{\Delta x}{\Delta t} = \frac{final\ height - initial\ height}{time}$$

22

• Statistical Analysis

The use of statistical test in Biology is an excellent tool because it confirms or rejects any correlation that may be forward by the candidate, in a scientifically reliable way. Please note that in Biology we use a confidence level 95% and significance level of p=0.05. If you use an online calculator, this feature is built in the software so there is no need to have reference to it. Also, the statistical tests need to state a null and an alternate hypothesis:

H₀: there is no significance difference/correlation between x and y

Hₐ: there is significance difference/correlation between x and y

Be aware, however, that **correlation does not mean causation.** If you use an online calculator, this feature is built in the software so there is no need to have reference to it. Here are the most common tests used in Biology:

1. Student's t-test

This test evaluates whether there is a significant difference between two sets of data. You need a minimum of 5 measurements per set though. For instance, the t-test would be ideal to determine if there is significant difference between pre- and post- caffeine caffeine consumption heart rate.

$$t = \frac{(x_1 - x_2)}{\sqrt{\frac{S_1^2}{n_1} + \frac{S_2^2}{n_2}}}$$

x1: the mean of sample 1, x2: the mean of sample 2

s1: the standard deviation of sample 1, S2: the standard deviation of sample 2

n1: the size of sample 1, n2: the size of sample 2

Online t-test online calculator

2. Pearson's correlation

This test evaluates whether there is positive or negative correlation between the dependent and independent variable. The Pearson's correlation coefficient, **r**, is calculated and can get values between -1 and +1. If the r-value is close to -1 or +1, this is a proof of strong negative or positive correlation respectively. The closer the value is to zero, the weaker is the correlation. A scatter graph can also be produced by the same software, to show this relationship. You can use, GDC, MS Excel, Google sheets or the Pearson's online calculator.

$$r=\frac{\Sigma XY-\frac{(\Sigma X)(\Sigma Y)}{n}}{\sqrt{(\Sigma X^2-\frac{(\Sigma X)^2}{nx})(\Sigma Y^2-\frac{(\Sigma Y)^2}{ny})}}$$

ΣX: sum of X values, ΣY: sum of Y values, n=number of 'pairs' of data.

3. Spearman's correlation coefficient

You need a minimum of 10 sets of data. The data is ranked (this automatic if you use an online calculator) and compared:

$$r_s=1-\frac{6\Sigma D^2}{n(n^2-1)}$$

r_s= Spearman's rank correlation coefficient

D= differences between ranks

N= number of pairs of measurements

If r_s is equal or above the critical value (p=0.05), the null hypothesis is **rejected**, which means there is a **significant** correlation between IV and DV.

4. chi-square test (x^2) of independence

This test evaluates if there is significant association between two variables or if they are independent. There are two hypotheses stated here:

H$_0$: there is no significant association between x and y. So, X and Y are independent of each other.

H$_1$: there is significant association between x and y. So, X and Y are dependent of each other.

This test is ideal for investigations in Ecology and Genetics.

$$x^2=\Sigma\frac{(O-E)^2}{E}$$

O= observed frequencies

E= expected frequencies = $\frac{row\ total\times column\ total}{grand\ total}$

In order to draw a conclusion, you need to compare your x^2 value with the critical value at p=0.05 at the appropriate degrees of freedom:

degrees of freedom=(number of rows -1)x(number of columns-1)

If the calculated x^2 value is larger than the critical value, accept null hypothesis, there is no significant association between X and Y. If the calculated x^2 value is smaller than the critical value, reject null hypothesis, there is significant association between X and Y, which means X is dependent on each other.

5. ANOVA (Analysis of Variance)

Often you need to compare more than two sets of data, such as 5 or 10, so instead of running multiple t-tests you can run an ANOVA test. There are great <u>ANOVA online calculators</u>, and the advantage is that all data comes out in one number.

H₀: there is no significant difference between the groups of data. Observed differences may be attributed to random error during sampling.

Hₐ: there is significant difference between the groups of data. Observed differences may not be attributed to random error during sampling.

5.4 Graphs

The graphs must contain the processed data only. There is no marks for graphs that use raw data. On the x-axis place the IV and on the y axis place the calculated DV. The essential components of a good graph are:

- The title on the graph
- The labels of x and y axis
- The error bars
- The legends
- The equation and R^2 value (from Excel)
- The interpretation of the graph a s a short text under the Graph.

There are two types of graphs. The one that plots **numerical** data and the one that plots **categorical** data. For numerical data, always choose the **best -fit line**. For categorical data choose a bar chart or histogram. The **best fit line** should go through all error bars.

The graphs in Example 1 and Example 2 presented below come from different studies on *Lens culinaris.* Example 3 comes from a study on the enzymatic hydrolysis of lactose from lactase.

Example 1:

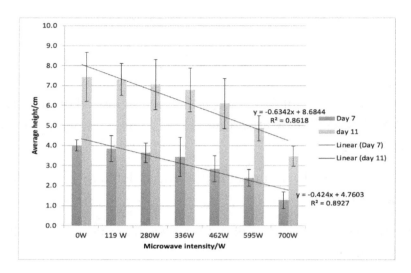

Interpretation of Graph

Line of best fit shows that there is a negative correlation as microwave intensity increases the height of L.culinaris decreases. The equation describing the trend line and the coefficient constant R^2 are shown in the graph. R^2 is not close to 1, so the correlation is moderate.

Examiner's comment: the title of the graph is missing. The best fit line is properly drawn with error bars and the equation is included. A conclusion is described, without interpretation however. It would be appropriate to calculate the % change in height for the x-axis.

Example 2:

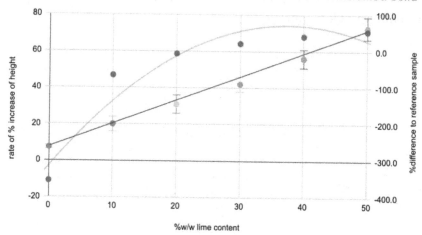

Rate of %increase of height of Lens culinaris seedlings and % difference to reference sample vs. %w/w lime content of acidified solid

The relationship between %w/w lime content and rate of growth is described from the linear equation, obtained from the graph above, $y = 1.3x + 5.22$ with $R^2 = 0.991$. There is a strong positive correlation between %w/w lime content and rate of %increase of height. This can be attributed to the fact that acidic pH affects the pectin and lignin content of the xylem vascular tissue of the roots, stalling the Cation Exchange Capacity, which is the primary mechanism for retrieving nutrients from the soil. However, liming, increases the pH, which further blocks toxic aluminium, and enables the CEC, therefore, the availability of important nutrients.

Examiner's comment: The graph has title, proper axis labels and an efficient interpretation. The student has not included the legends though. It is interesting that they included a comparative trendline that shows the %difference from the control. The student has not commented on that though.

Example 3:

Absorbance of ONP Product in Different Concentrations of Lactose

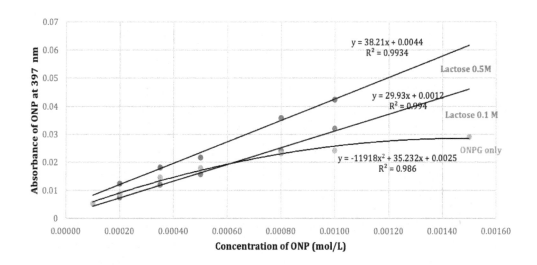

The trends seen from Tables 2 through 5 and this graph are that ONP production increases linearly with increasing concentration of ONPG, until it reaches a plateau. The role of galactose as a product of hydrolysis and its role as competitive inhibitor can explain the plateau as further analyzed in the discussion. The absorbance and concentration of ONP is decreased when we compare 0.1 M with 0.05 M lactose and both compared to the values obtained from the only-ONPG substrate. The error bars are inconspicuous.

Examiner's comment: the graph satisfies all criteria. It has a title, properly labelled axis and legends. It includes the equation, R^2 value and an interpretation. There is reference to the tables used for the interpretation. The error bars were also included but the student clarifies that they were too small.

---

6. EVALUATION

The Evaluation is a significant component of your IA. It should include the following:

- The conclusion that is justified based on appropriate scientific background
- A detailed and thorough discussion on the limitations of the investigation
- The suggestion of viable improvements and extensions for future investigations

6.1 Conclusion

In order to write an excellent conclusion, make a bullet point list of all the questions that the conclusion show refer to. Here is a suggested flow:

- What was the aim of the investigation initially and/or the hypothesis?
- Has the data processing answered the research question?
- What is the numerical data you have obtained? Refer to the high and low values, the trends, the correlations and the graphs. Make use as much as possible of the data you have acquired and never forget to include the quantitative data as well.
- Compare the findings of your investigation with other researchers'. Make a clear reference to one or two studies and evaluate how the information from those studies could interpret the findings in your IA.
- Recap on the significance of your study, as you mentioned in the background theory. How your findings and the conclusion can be related to real life?

The example below are the last two paragraphs of a Biology IA that studied the thermal line of *Phaseolus vulgaris*:

Example:

> The findings are supported by the fact that the enzymes involved in the biochemical processes of germination, have optimum temperatures between 20^{0} C and 35^{0} C. So, colder temperatures slow down cellular respiration and ATP production, hence, the effect is that the growth is stalled as energy is insufficient. At higher temperatures, namely at 40^{0} C, the enzymes are denatured so the processes of respiration and germination slow down. In effect in the present investigation, although $120^{0}C$ is quite high, the amount of time the seeds are exposed seems to be sufficient for conditions to enhance germination but more time of exposure is detrimental and denatures the enzymes.
>
> Nakao and Cardoso (2016), in a similar study, in which they developed a thermal time model for Urochloa sp.; they concluded that, indeed there is an optimum temperature combined with optimum time, although the model would not suffice for a wider range of temperature-time combinations. Also, it is claimed that this model can explain the rapid germination of many tropical and mediterranean plants. However, each individual species has its own thermal line assay. In *Phaseolus vulgaris*, in particular, the understanding of its thermal line will empower the sustainable cultivation of this species, which has enormous economic impact on agricultural economy, so that the maximum yield is obtained.

6.2 Limitations and improvements

Both strengths and limitations should be addressed. Always, mention one or two strengths of your study with justification.

The limitations must be discussed on two levels. Firstly, explain how the identified limitation has possibly affected the investigation. What is the scientific background of this disturbance? Does it create a random or a systematic error? Secondly, offer a viable and meaningful improvement. Make sure the suggestions are not superficial or naive.

There should be at least 5 limitations addressed.

Here is an example of a part of a limitations' table from a study for the effect of pH on the growth of lentils:

Example:

The distribution of water into the clay soil is assumed to be even and water capacity is assumed to be reached in all samples	Clay soil has pores of different sizes and may hold water but the rate with which water is filling up the pores between samples may not be even providing different availability of water to the growing lentils, skewing the results for the rate of growth.	The water retention tests are tedious and complex for a school setting. However, the clay soil mass in the pot used could be measured before and after watering to record the amount of water change. Alternatively, the volume water drained under the pot could be daily monitored.	**Examiner's comment:** There is satisfactory reference to these two limitations with proper justification. There should be some reference as to whether the error is random or systematic.
The ability of the xylem to carry out CEC was not evaluated.	The reduction of pectin or lignin due to low pH is not always consistent to all samples exposed to low pH, which may affect the CEC and the therefore the growth rate.	Microscopic slides of the root xylem may provide additional information and evaluate the uniformity of the deterioration of the xylem.	

6.3 Extensions

The extensions should not be the same as the suggestions for improvement, which is a common mistake. Do some research and evaluate whether there is groundbreaking research on the topic you chose. What bout looking into the iV and DV from a different perspective?

7. COMMUNICATION

The criterion of Communication is also holistic, which means it is marked throughout the IA report. In order to get full marks for Communication:

- The lab report should be comprehensive and flow. The reader should be able to grasop the aim of the investigation and the research question early on.
- There is appropriate use of scientific vocabulary. Avoid simple words to describe a biological process. The organisms should be mentioned with their scientific name. For instance the phrase 'the sucrose molecules diffuse along the semipermeable membrane along a concentration gradient' is more appropriate than 'the sugar molecules move to the other side'.
- The presentation of the tables and the graphs as has been thoroughly discussed previously in the Analysis section: clear headings, units with uncertainties, labels, centered columns, consistency in significant figures and decimal places.
- The bibliography should be included as in-text citation in the main body and as full citation at the end of the report. You may use any of the referencing systems applicable to IB such as MLA, APA, Harvard etc. There are quite a few citation machines such as citation machine and easybib.

PART II
SEVEN EXAMPLES OF EXCELLENT INTERNAL ASSESSMENT

The assessments featured in this section are all recently submitted IAs that scored very highly after being moderated by the IBO. To prevent plagiarism and duplication of results, the appendices have been omitted. The IAs are presented in the exact same way as they were submitted, and without any edits or changes to formatting.

We do not retain the copyright of these commentaries, nor is this publication endorsed by the IBO. The Internal Assessments are being re-printed with the permission of the original authors.

1. THE EFFECT OF VARYING INITIAL CONCENTRATIONS (M) OF COPPER (II) SULFATE SOLUTION ON THE CHANGE IN CONCENTRATION OF THE FINAL SOLUTION

Author: Adam Zhou
Moderated Mark: 19/20

1. Introduction

Eichhornia crassipes, or the common water hyacinth, though endemic to the Amazon basin, has been introduced in many different tropical ecosystems as an invasive species (Villamagna, et. al. 2010), such as that of the Philippines, where I was born and raised. Environmentalists have long been concerned about its presence since it floats on water bodies and grows at a fast rate, it lowers light penetration to organisms living underneath, hindering photosynthesis, and absorbs dissolved oxygen that these other organisms need to survive. Yet, I believe that due to its ability to absorb heavy metals dissolved in water, especially that of copper, lead, and manganese, the excess water hyacinth can be used for bioremediation purposes when redirected towards polluted water bodies (Fu, et. al., 2015). This specific type of bioremediation is known as phyto-accumulation or phyto-extraction, involves the roots absorbing the contaminants along with the other nutrients and the surrounding medium of water. The roots have a large surface area to volume ratio which allows for more absorption from the contaminated microorganisms and soil by means of the xylem loading (Köhler, 2007). Xylem loading takes place at the xylem parenchyma root, where the cell walls of the vascular tissue act as apoplastic paths for the free absorption of nutrient circulation (Etim, 2012). These three components (i.e. roots, microorganisms, and soil) undergo a symbiotic relationship, where the roots release exudates into the rhizosphere (the narrow region of soil known as the root microbiome), and thus affects the prolific activity of microorganisms while maintaining compaction and quality of soil. In turn, the microorganisms and soil will aid in the maintenance of the amount of contaminant that the plant will be able to absorb (Ramachandra, 2004). This absorption will be done through the presence of transporters, which are transmembrane proteins in the semi permeable membrane of the rhizosphere.

Given increased rates of urbanization and the functionality of factories, improper disposal of chemicals into water bodies has also seen an increase. If such rates continue, toxic levels may fare too high for the water hyacinth to be able to still absorb ions, and hence, this experiment tests for the efficacy of this plant in higher than normal concentrations. Thus, this paper aims to identify: **What is the effect of varying initial concentrations (M) of copper (II) sulfate solution on the change in concentration of the final solution based on the amount of absorption of *Eichhornia crassipes* measured by spectrophotometer analysis over a period of 12 hours?** I chose this topic with my background and interest in environmental sustainability research. In being a national representative for World Oceans' Day as well as an ambassador for Samsung Engineering's environmental networking program, I have been exposed to the natural resource degradation from sites of wonder to those of desolation. I've also learned how freshwater, though constituting 1% of the world's surface, houses a quarter of all known vertebrates ("Freshwater Biodiversity") as well as all of human domestic consumption. The interconnectedness of the human and ecological systems posits the need for sustainable practices, such as managing the presence of phyto-accumulating organisms.

2. Investigation

2.1. Hypothesis

Given a period of 12 hours, the water hyacinth will be able to absorb copper (II) sulfate from the solution, for all the different concentrations tested. A greater concentration, however, will yield lesser absolute absorption of the solution, assuming toxic build-up in the plant stem and roots. This assertion is the H_1 or the alternative hypothesis which states that there is a significant difference between initial concentration of copper (II) sulfate solution on the amount of absorption of the water hyacinth. Then, the H_0 or null hypothesis value is that there is no significant difference between these two variables.

2.2. Variables

Independent Variable: Initial concentration of copper (II) sulfate solution (M)
Dependent Variable: Change in concentration of copper (II) sulfate solution (M)
Controlled Variables:

- The volume of the solution for the water hyacinth. Due to evaporation rates that arise especially due to high temperatures and direct sunlight, the same concentration of solution is added to compensate for such. A marking on the tub has been made. In addition, with readings for the spectrophotometer necessitating the removal of 3 mL of the solution into a cuvette, the 3 mL will be replenished inside the tub.
- The time of day when readings were taken. This ensures that the time in between measurements is exactly 2 hours only and thus, no excess absorption will take place. This will be done over a 12 hour period.
- The concentration of the copper (II) sulfate solution.
- All water hyacinths were sourced from Jefferson Capati's Aquarium store so that the abiotic conditions they were in beforehand were as similar as possible.
- The species of water hyacinth. Different plants may have different absorption rates and minute differences in stem structures, thus the species is only *Eichhornia crassipes*.
- Same level of light source. Though this could not have been maintained at a steady rate due to the differences in sun exposure day to day, and the occurrence of unexpected weather events, the plants were placed in the same location of the windowsill as sunlight exposure has a positive correlation with plant metabolism activities.
- Same general mass of water hyacinth. The larger the plant has more roots which can allow for more absorption, and though this cannot be kept exactly constant due to the nature of plant diversity, data processing will consider per-gram concentration difference.

2.3. Preliminary Experiment

The two days prior to the actual experiment were devoted to finding necessary adjustments to the methodology if such were needed. Though the initial perceived time frame needed for this experiment was a week to allow for full absorption, it was found that at 12 hours after the experiment has been conducted, the water hyacinth had died and that the level of absorption of the solution by spectrophotometric analysis plateaued. In addition, the midpoint concentration of copper (II) sulfate was adjusted from 0.02 M to 0.04 M to allow for the absorption levels measured in the spectrophotometer to have readable values and have a wider scope of allowance for the water hyacinth to absorb the solution. 0.06 M was also tested as the highest mark for an independent variable but resulted in the water hyacinth dying too quickly.

3. Procedure

3.1. Apparatus:
- Anhydrous copper (II) sulfate
- Deionized water
- Metal spatula
- 25 small transparent tubs (7.5cm radius, 9cm height)
- 25 water hyacinth plants
- Graduated Cylinder 100 mL volume ($\pm$ 0.5mL)
- Vernier Go Direct™ Spectrovis Plus Spectrophotometer
- Laptop with Logger Pro Software
- 5 Spectrophotometer Cuvettes
- 5 Plastic Pipettes ($\pm$ 0.5 mL)
- Mass balance ($\pm$ 0.01g)
- Stirring rod

3.2. Setup

Figure 1: Photograph of Setup

Photograph taken by myself

3.2. Methodology

After preparing the appropriate mass needed to obtain a specific concentration of copper (II) sulfate in a 3L volume, place 500 mL into each of the 5 small transparent tubs, measured by a graduated cylinder. Now, with a mass balance, measure the mass of the water hyacinth and place

38

into solution, taking note to label the tub accordingly to mass and initial concentration. Every two hours, a sample of around 3 mL will be extracted to fill a cuvette for the spectrophotometer. Note should be made to stir solution to dissolve any precipitate, if any, beforehand, while refilling the tub with 3 mL from excess supply of the same concentration. Cuvettes will now be inserted, with the transparent side facing the light source, into the spectrophotometer which is linked into a laptop with Logger Pro software. Calibration will ensue to obtain the wavelength used by the spectrophotometer. Afterwards, the screen should now indicate the absorbance level. Repeat these steps for the other independent variables.

3.3. Justification

Copper sulfate is a solution that constitutes the toxic copper metal, found in polluted waters next to factories and agricultural areas. Especially since it is a naturally occurring pesticide, its commercial application in the development of fertilizer mixtures and as a reagent in chemical engineering also makes it a common heavy metal, thus relating to the real life context of this investigation. The concentrations used, ranging from 0.02 to 0.06 M is an above average toxic concentration that emulates a near future projection of contaminated waters ("Wastewater Characteristics and Effluent Quality Parameters"). Meanwhile, the deionized water it was dissolved in was chosen to ensure no minerals would react with the metal.

Having an extensive number of time intervals acting as the independent variable allowed for a more accurate depiction of the data collected and how this trend occurs over a period of time. The dependent variable is the absorbance of light through a given solution, that of which will be analyzed to calculate the concentration of the remaining solution. The change concentration is then an observation of the loss in metal ions.

3.4. Safety, Ethical, and Environmental Implications

Safety: The copper (II) sulfate is, according to the Material Safety Data Sheets (MSDS), a toxic substance and is a known irritant, thus safety goggles and gloves will be worn when conducting its measurement or pouring. Sections V and VI of the MSDS states that if in contact with skin, the affected area will be washed with soap immediately. If inhaled, swallowed, or in contact in the eyes, a physician should be contacted.
Ethical: The obtaining of the water hyacinth may have interactions with consumers in a given area, either as a food source or acting as the niche of a given species. As it is necessary for the experiment, minimal amounts of such will be sourced.
Environmental: The disposal of the water hyacinth and the solution must be done properly. Since both of these have nutrients which can promote eutrophication, they will be diluted with water to reduce its toxicity and the hyacinth may act as fertilizer.

4. Data Presentation

Table 1: Raw Data Table Showing the Absorption of copper (II) sulfate Solution by Spectrophotometer Analysis Over a Period of 12 hours with 2 hour intervals

Initial Concentration (M)	Trial #	Mass (g ± 0.01)	Absorption of copper (II) sulfate Solution at 949.9 nm wavelength (Au ± 0.01)						
			0:00	2:00	4:00	6:00	8:00	10:00	12:00
0.02	1	15.97	0.320	0.271	0.229	0.201	0.188	0.179	0.173
	2	18.83	0.320	0.265	0.224	0.205	0.189	0.176	0.169
	3	16.82	0.320	0.269	0.220	0.206	0.192	0.178	0.170
	4	17.06	0.320	0.276	0.232	0.192	0.190	0.175	0.170
	5	15.80	0.320	0.268	0.231	0.209	0.197	0.180	0.174
0.03	1	15.01	0.353	0.250	0.231	0.205	0.191	0.185	0.182
	2	15.92	0.353	0.248	0.233	0.211	0.192	0.189	0.186
	3	13.86	0.353	0.255	0.230	0.208	0.190	0.185	0.181
	4	14.40	0.353	0.261	0.229	0.220	0.197	0.186	0.182
	5	17.89	0.353	0.247	0.222	0.198	0.184	0.179	0.174
0.04	1	15.13	0.599	0.428	0.420	0.413	0.404	0.402	0.399
	2	16.49	0.599	0.419	0.412	0.408	0.394	0.392	0.391
	3	14.22	0.599	0.422	0.415	0.406	0.404	0.404	0.403
	4	15.71	0.599	0.425	0.419	0.411	0.402	0.399	0.396
	5	14.43	0.599	0.421	0.409	0.400	0.396	0.394	0.393
0.05	1	14.44	0.633	0.431	0.420	0.418	0.416	0.411	0.413
	2	12.90	0.633	0.445	0.438	0.432	0.425	0.419	0.418
	3	13.89	0.633	0.435	0.431	0.427	0.421	0.414	0.415
	4	14.01	0.633	0.429	0.425	0.420	0.409	0.407	0.409
	5	13.53	0.633	0.447	0.438	0.433	0.420	0.414	0.414
0.06	1	13.24	0.759	0.513	0.500	0.487	0.482	0.481	0.480
	2	16.33	0.759	0.520	0.501	0.496	0.494	0.492	0.489
	3	16.62	0.759	0.522	0.514	0.501	0.489	0.488	0.488
	4	15.86	0.759	0.519	0.492	0.486	0.483	0.479	0.481
	5	13.57	0.759	0.511	0.503	0.497	0.493	0.491	0.490

During the experiment with data collected above, it was noted that in the first hour, there were already noticeable differences, mainly in the color of the solution remaining. Furthermore, the roots of the water hyacinth started to turn a faint blue. At around the fourth hour, yellowing and wilting of the plant started to occur, where by the eighth hour, this qualitative observation was severe, to turn slightly brown. By the tenth hour, spots of blue showed up on the plant, and appeared almost dead, with no noticeable difference by the twelfth hour.

To find the concentration based off absorption level, we need to apply Beer-Lambert's Law:

$$A = \varepsilon l c$$

where A = absorption, ε = molar absorption coefficient of the solution, l = length of the solution the light passes through, and c refers to the concentration of the solution in Molarity (M) ("The Beer-Lambert Law").

Rather than calculating and substituting the variables, a more direct application of this law is in how it states that absorbance is directly proportional to concentration of the solution. Thus, one can use a spectrophotometer which measures the amount of light absorbed by passing through a solution to its concentration. After identifying the absorbance levels of five different concentrations of a given solution, in this case, copper sulfate, and applying a linear regression to get the equation of a line, the points of unknown concentration and known absorbance will follow this trendline. Hence, substituting the y value of absorbance into this equation will yield the x value, that is, of concentration.

Using the time 0:00 for each of the different known concentrations of the solution, 5 points will be obtained on a trendline once absorption is found with a spectrophotometer. These are given in the following figure below.

Figure 2: Photograph of Samples of Copper Solution (II) Obtained in Cuvette in Varying Concentrations to Test for Beer-Lambert's Law

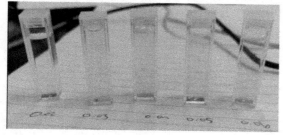

Photograph taken by myself

Using Logger Pro, the y = mx + b regression line can be obtained.

$$y = 11.58x + 0.0696$$

41

Substituting initial and final absorption values from Table 1 is now shown in Table 2 below:

Table 2: Processed Data Table Showing the Concentration of copper (II) sulfate Solution by Beer-Lambert's Law Over a Period of 12 hours with 2 hour intervals

Initial Concentration (M)	Trial #	Mass (g ± 0.01)	Concentration of copper (II) sulfate Solution (M ± 0.01)						
			0:00	2:00	4:00	6:00	8:00	10:00	12:00
	1	15.97	0.0216	0.0174	0.0138	0.0113	0.0102	0.0094	0.0089
	2	18.83	0.0216	0.0169	0.0133	0.0117	0.0103	0.0092	0.0086
	3	16.82	0.0216	0.0172	0.0130	0.0118	0.0106	0.0094	0.0087
	4	17.06	0.0216	0.0178	0.0140	0.0106	0.0104	0.0091	0.0087
0.02	5	15.80	0.0216	0.0171	0.0139	0.0120	0.0110	0.0095	0.0090
	1	15.01	0.0245	0.0156	0.0139	0.0117	0.0105	0.0100	0.0097
	2	15.92	0.0245	0.0154	0.0141	0.0122	0.0106	0.0103	0.0101
	3	13.86	0.0245	0.0160	0.0139	0.0120	0.0104	0.0100	0.0096
	4	14.40	0.0245	0.0165	0.0138	0.0130	0.0110	0.0101	0.0097
0.03	5	17.89	0.0245	0.0153	0.0132	0.0111	0.0099	0.0094	0.0090
	1	15.13	0.0457	0.0309	0.0303	0.0297	0.0289	0.0287	0.0284
	2	16.49	0.0457	0.0302	0.0296	0.0292	0.0280	0.0278	0.0278
	3	14.22	0.0457	0.0304	0.0298	0.0291	0.0289	0.0289	0.0288
	4	15.71	0.0457	0.0307	0.0302	0.0295	0.0287	0.0284	0.0282
0.04	5	14.43	0.0457	0.0303	0.0293	0.0285	0.0282	0.0280	0.0279
	1	14.44	0.0487	0.0312	0.0303	0.0301	0.0299	0.0295	0.0297
	2	12.90	0.0487	0.0324	0.0318	0.0313	0.0307	0.0302	0.0301
	3	13.89	0.0487	0.0316	0.0312	0.0309	0.0303	0.0297	0.0298
	4	14.01	0.0487	0.0310	0.0307	0.0303	0.0293	0.0291	0.0293
0.05	5	13.53	0.0487	0.0326	0.0318	0.0314	0.0303	0.0297	0.0297
	1	13.24	0.0595	0.0383	0.0372	0.0360	0.0356	0.0355	0.0354
	2	16.33	0.0595	0.0389	0.0373	0.0368	0.0366	0.0365	0.0362
	3	16.62	0.0595	0.0391	0.0384	0.0373	0.0362	0.0361	0.0361
	4	15.86	0.0595	0.0388	0.0365	0.0360	0.0357	0.0354	0.0355
0.06	5	13.57	0.0595	0.0381	0.0374	0.0369	0.0366	0.0364	0.0363

Table 3: Processed Data Table Showing the Change in Concentration of copper (II) sulfate Solution, with Analysis per gram of plant tissue by Spectrophotometer Analysis, Including Average and Standard Deviation

Initial Concentration (M)	Trial #	Mass (g ± 0.01)	Change in Concentration over 12 hours (M ± 0.01)	Change in Concentration per gram of plant tissue (M ± 0.01)	Average Change in Concentration per gram of plant tissue (M ± 0.01)	Standard Deviation Change in Concentration per gram of plant tissue (M ± 0.01)
	1	15.97	-0.0127	-0.000795		
	2	18.83	-0.0130	-0.000692		
	3	16.82	-0.0130	-0.000770		
	4	17.06	-0.0130	-0.000759		
0.02	5	15.80	-0.0126	-0.000798	-0.000763	0.000042
	1	15.01	-0.0148	-0.000984		
	2	15.92	-0.0144	-0.000906		
	3	13.86	-0.0149	-0.00107		
	4	14.40	-0.0148	-0.00103		
0.03	5	17.89	-0.0155	-0.000864	-0.000970	0.000085
	1	15.13	-0.0173	-0.00114		
	2	16.49	-0.0180	-0.00109		
	3	14.22	-0.0169	-0.00119		
	4	15.71	-0.0175	-0.00112		
0.04	5	14.43	-0.0178	-0.00123	-0.001150	0.000058
	1	14.44	-0.0190	-0.00132		
	2	12.90	-0.0186	-0.00144		
	3	13.89	-0.0188	-0.00136		
	4	14.01	-0.0193	-0.00138		
0.05	5	13.53	-0.0189	-0.00140	-0.001380	0.000046
	1	13.24	-0.0241	-0.00182		
	2	16.33	-0.0233	-0.00143		
	3	16.62	-0.0234	-0.00141		
	4	15.86	-0.0240	-0.00151		
0.06	5	13.57	-0.0232	-0.00171	-0.001580	0.000182

To calculate for change in concentration, apply the following formula:

$$Concentration\ at\ Hour\ 12\ -\ Concentration\ at\ Hour\ 0$$

E.g. for 0.01 M Initial Concentration Independent Variable, Trial 1:
$0.0089 - 0.0216 = -0.0127\,M$

To calculate for change in concentration per gram of plant tissue, apply the following formula:

$$\frac{Concentration\ at\ Hour\ 12 - Concentration\ at\ Hour\ 1}{Mass\ of\ Plant\ Tissue}$$

E.g. for 0.01 M Initial Concentration Independent Variable, Trial 1:
$\frac{0.0089 - 0.0216}{15.97} = -0.000795\ M$

To calculate for average concentration per gram tissue value, apply the following formula:

$$\frac{\sum Concentration\ of\ Each\ Test\ Trial}{Number\ of\ Test\ Trials}$$

E.g. for 0.01 M Initial Concentration Independent Variable
$\frac{-0.000795 - 0.000692 - 0.000770 - 0.000759 - 0.000798}{5} = -0.000763\ M$

Now, a visual representation of the processed data is displayed in the following:

Figure 1: Scatter Plot Showing the Relationship Between Initial Concentration of Copper (II) Sulfate on the Change in Concentration of Copper (II) Sulfate after 12 Hours Based on the Amount of Absorption of *Eichhornia crassipes*, where Error Bars Indicate Standard Deviation

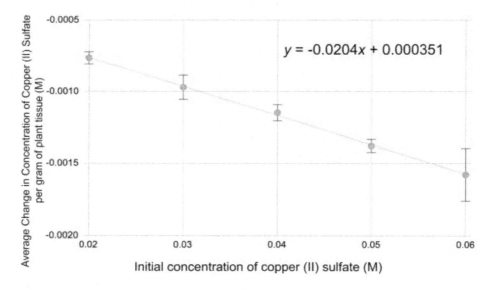

As seen in Figure 1, there is an inversely proportional relationship between the initial concentration of copper (II) sulfate solution and the change of concentration after a twelve hour period, with a steady linear decrease. No anomalies are seen, though the initial concentration of 0.06 M has a noticeably larger error bar, which is represented by standard deviation of the sample.

Now, to determine correlation with an assumption that the test sample is normally distributed, the Pearson's Moment Coefficient (r^2) from Excel software is used for a statistical test (McDonald, 2009). This calculated value is -0.99. When considering the range of this r^2 value from -1 to +1, it will be deduced that -0.99 is a near perfect negative correlation between the independent and dependent variables. Comparing this value to the levels of significance table, and noting how the degrees of freedom is 3, under a 0.05 significance level, a ±0.878 r^2 value is at least needed and under a 0.01 significance level, a ±0.959 r^2 value is needed. Therefore, with both levels, there is a significant result obtained. This value also highlights how there is a low chance of error and thus a greater accuracy of the experimental procedure, shown also in how the data point and the linear regression line are lined up with each other.

5. Conclusion

Based on the results obtained from this experiment, it can be determined that as the initial concentration of copper (II) sulfate solution that a given organism is subjected to increases, the capability of the organism to absorb such toxins, as measured by the change in concentration of the overall solution, decreases. With this conclusion, the null hypothesis stated in the aforementioned is accepted and the alternative hypothesis is rejected. Yet, it is noted that there is still a significant difference between the independent and dependent variables, just in terms of the opposite of the hypothesized result. This pattern could be justified given the properties of osmoregulation, and more specifically osmotic pressure. When the solvent molecules of the toxins are moved from a low-concentration solution to a high-concentration one, an equilibrium must be instituted as to ensure proper metabolic functions ("Osmotic Pressure"). If the initial concentration is too heavy, these metabolic conditions are to be affected, and will thus have less amount of absorption. This pattern is seen where the amount of absorption per gram of tissue in terms of concentration change of the copper (II) sulfate solution was significantly lower with an initial concentration measured at 0.02 M compared to 0.06 M with results at -0.000763 M and -0.001580 M changes respectively.

For instance, this effect is seen as the toxicity is accumulated in plants, this experiment was able to show this given how the manifestation of chlorosis or yellowing of the foliage is present in this experiment. This is mentioned in the qualitative observations of section 4. This biological phenomenon is caused when excessive amounts of copper interferes with the iron metabolism in a flora organism (Etim, 2012). Once this hyperaccumulation of a toxin is introduced, the channels in the rhizosphere close, hindering the uptake done by protein transporters in the semi-permeable membrane. This is another explanation for the trend seen in the experiment (Ramachandra, 2004).

6. Evaluation

Strengths:

Given how the experiment used apparatus with low uncertainties, the overall systematic error was low as well. This, factored in with the consistency in following the procedure including its associated control variables. Conducting a wide range of both intervals in the independent variables as well as the trials per each of them helped give a wide range of data. This determination of whether there is consistency or not helped determine the validity of the conclusion and offer a more accurate representation of such.

Weaknesses:

From qualitative observations, the precipitation of the anhydrous copper (II) sulfate rather than being full dissolved could lead to an improper reading by the spectrophotometer. This is since the cuvette only takes an aliquot from a place that may or may not be more concentrated than what it is actually meant to be given if that area has more precipitate. Though dissolving was taken note of in the procedure, it was not always managed to be fully dissolved back. For future references, more time should be spent stirring the solution no matter the inconvenience.

There is also the assumption made that each gram of tissue is representative of a tissue that is responsible for the absorption of the toxins. Rather this mass could be present in a more abundant foliage with little fibrous root stems, or vice versa. This would lead to inaccurate measurements on whether a plant is able to absorb more toxins or not. For future experiments, note could be made to collect similarly sized water hyacinths or even at the end of the experiment, cut off the root system parts for appropriate measurement.

The evaporation of the copper (II) sulfate solution into the atmosphere was not accounted for and with a decrease in volume, leads to a variation in how the system was set up in the initial parts of the experiment. As more time passes by, and the volume decreases, there may be a plateau in the amount of solvent the water hyacinth was able to absorb. Thus the plant could be set up in a beaker like tub with volume measurements indicated. Note would be made in every time interval to restore the volume to its initial state whilst considering the same concentration.

There has always been a variation in the reading of the spectrophotometer with no set value being given. Though this makes the collected value to be placed into the data table unclear, a measurement of all the values the spectrophotometer gives after a minute long waiting period to allow for adjustment time, and then its average will allow for a more accurate reading. Furthermore, one can introduce more trials to discount the variations.

An inherent flaw in the methodology lies in how it tests for higher than what is expected concentration than what is present in current wastewater levels, but rather takes into account a future projection. This had to be done especially considering the scope of the spectrophotometer's measurements, however, would not be an accurate representation of the real world and could be more properly adjusted to fit. In accordance with such, a titration method would allow for a wider range of independent variable subjects.

Similarly, another inherent flaw in this investigation is how the data obtained takes into consideration more of short-term impacts given the time frame given the 12 hour set up allotted for data collection. Though the data only allowed for such given how the plant has noted symptoms of chlorosis, in fixing the previous issue with an appropriate independent variable range, a lower alternative would yield a more long-term possibility for data collection.

The lack of sterilization of equipment used in the measurement and weighing of chemicals as well as minimal sterilization of the plant itself could have introduced other toxins that would interfere against the metabolic pathways, either accentuating the results outlined in the conclusion with more toxins hindering the levels of absorption of water hyacinths, or also potentially having unwanted chemical reactions. To counter this, a chelating agent such as EDTA could be used, or more simply, involve the use of proper sterilization.

7. Extension

For further study, one can observe a wider variety of heavy metals such as copper, lead, and manganese, and the ability of water hyacinth to still absorb these compounds. This will allow an analysis on the versatility of the water hyacinth for various concentrations which is more feasible in the real life application on the contamination of water bodies from factories. Similarly, it would be interesting in testing other plants that have a phyto-accumulation characteristics such as shrub tobacco (*Nicotiana glauca*), and members of the *Arabidopsis* genus. In a more holistic sense, one can also identify the genetic source for such plants' phytoremediation. Given how the selenocysteine methyltransferase (SMTA) gene from the selenium hyperaccumulator has been used as transgenic tool in artificial selection, such principles can be modeled over as an extension to this experiment and see which genes are most effective in the aforementioned role (Etim, 2012).

2. THE EFFECT OF DIFFERENT ELECTRICAL VOLTAGES ON THE GROWTH OF VIGNA RADIATA

Author: Bui Philong
Moderated Mark: 22/24

The effect of different electrical voltages on the growth of *Vigna radiata*

Background

According to Collins dictionary (n.d.), electro culture is the practice of using electricity in agriculture in order to stimulate plant growth. Plants do not have a nervous system for electricity to pass through; however, electricity affect its growth and biological properties (Nytimes.com, 1985). There have been studies showing how manipulating electric current around plants, could boost their growth. This could help in agriculture preferences, as an alternative to using fertilizers or when a growing season is limited, to boost the growth of crops. Electrical charges are used as transport of materials in and out of cells, and they further regulate metabolism in cells (Jeanty, n.d.). The increase in metabolism would increase the ion pumping transport of hydrogen ions inside the cells increasing the nutrition value of the plant, additionally, the electricity would also affect and increase the water uptake of plants (What-when-how, n.d.). The flow of electrical surges passing through the plant has the potential of manipulating and stimulating the increase in the nutritional uptake for plants, and thus, affecting the growth of plants. This experiment would expose the plant to different electrical voltages.

Statement of purpose

I was intrigued to do this experiment after I saw my dad leaving a charger by accident in a pot of plants. I was curious how the electrical stimulation could have affected the plants. It is quite common for outdoor wires to be damaged exposing their bare-naked electrical cables to the ground. Furthermore, the bare wires could send electrical surges to the soil affecting the plants around it. As the world is developing with technological advances and increasing electrical wires filling the planet, this research could convey on how bare electrical wires in the industrial world would affect the plants or in the practice of electro culture to help in agriculture growth.

Figure 1. Complex apparatus

A complex apparatus had to be made for the experiment, as seen in Figure 1, involving a parallel circuit around each group of plants with two electrodes inserted into the soil of each pot. Direct current (DC), an electrical charge would flow from one rode into the soil and plant and then back into the second rode. This would allow the soil to get electrically charged and potentially allow the plants to be exposed to electrical surges with different voltages. For the experiment, *Vigna radiata* (also known as mung bean), from the family of Leguminosae (bean plants family) has been selected due to the fast-growing natural properties of the plant. The experiment would use pre-germinated seeds, which ensures that all the plant is alive by imbibing the seeds in water, to allow absorption of liquid and trigger the seeds germination before the experiment.

Additionally, ammonium sulfate, $(NH_4)_2SO_4$ would be added to the soil, which will act as a fertilizer, but mainly used to allow electricity to conduct through the soil. As growth is described as the increase in size, the height and mass would be measured. Finally, as it is unsure whether the electrical stimulation would make plants grow by cell division through more nutrition or just an increase in water uptake; the wet mass, and dry mass would be measured to check the water content of the plants.

Aim

This biology experiment aims to investigate the effects of different electrical voltages in soil on pregerminated seeds of *V. radiata* by measuring its growth through the height and the final mass of the plant.

Research Question

How does different electrical voltage (3V, 6V, 9V, 12V, 16V) affect the growth of pre-germinated seeds of *Vigna radiata* measured through their average height (mm), dry mass (mg) and wet mass (mg) over 20-day period experiment?

Hypothesis

Electrical voltage will successfully help boost the growth of *V. radiata* up to an optimum level. It will help boost the plants uptake of water and increase the metabolism allowing the plants to grow at a faster rate through their height and increase in mass (Jeanty, n.d.).

Prediction

It is predicted that up to an electrical voltage, the growth of the plant would be the highest with the highest average height and mass (higher than control group), however, higher than the optimum voltage level, the vitality of the plant will decrease as the heat from the electricity would kill and stunt the plant growth decreasing its growth.

Variables

Table 1: Groups for the experiment

Experimental Groups	3V, 6V, 9V, 12V, 16V	Each group with 18 seeds would have the same conditions, but only the voltage of the electrical surge would be manipulated. There is an increase of 3V per each group, to see the effect of increasing voltage per plant.
Control Groups	0V	Group 0V would have the same conditions as the experimental group, except it will not be exposed to any electrical surge. This would be the control group for the experiment.

Table 2: Independent and Dependent Variables

Independent variable	Different level of voltages passing through the soil: 3V, 6V, 9V, 12V, and 16V	This was controlled with a power supply by turning the voltage switch. Model "MCP-m10-sp303e-DC" and "Frederiksen-DC" power supply was used. Each pot in each group was ensured the same voltage due to the parallel circuit and nails attached to the pots.
Dependent variable	The growth of *V. radiata* measured through: -Height every 2 days (mm) -Wet mass (mg) measured on day 20 -Dry mass (mg) measured on day 20	The height was measured using the same ruler for the whole experiment with a ± 5mm uncertainty. The mass was weight with the same digital scale all the time with a ±100mg uncertainty.

Table 3: Controlled variables and method of control

	Reason for it to be controlled	Method of control
Soil	Soil provides adequate nutrients and foundation base for the plants to grow in optimal conditions.	All the pots received the same amount of soil of 130g (weight on a digital scale) from the same company which provided 100% natural organic soil for growing crops.
Water content during watering [ml]	To make sure that all the pots receive an equal amount of water, as water it one of the factors which affect plant growth	With a measuring beaker, 20ml of water was measured and poured in every pot
Temperature [°C]	Temperature is a factor affecting enzyme activity. It is also a limiting factor, as after an optimum level, it decreases the rate of photosynthesis of *V. Radiata*	All the plants were arranged and placed in the same closed room with a air-conditioning which regulated the temperature so they were all exposed to the same temperature of 23-25°C. It was monitored using the air conditioning.
Humidity	Water vapor affects the transpiration and respiration of plants, by opening the stomata (Polygon, n.d.)	All the plants were arranged in the same enclosed room with the air-condition regulating and measuring the same humidity rate of 40% RH. This ensured that all the plants would be surrounded with the same humidity rate equally.

Pre-germinated seeds	With the seeds already pre-germinated in the experiment, this would ensure that the seeds are in a vital state and growing before being exposed to the treatment.	In order to make all the seeds germinated, they were imbibed in a bowl of water. After that, visually selecting the best seeds with a similar growing radicle ensured that the seeds are alive and starting to grow at the same time period.
Sunlight [lux]	Sunlight intensity affects the Carbon dioxide uptake, and thus the rate of photosynthesis	The plants were positioned in an enclosed balcony, where the walls and roof were made of transparent glass. This allowed the sunlight to spread equally among the plants without shadows caused by concrete walls. This was monitored with a smartphone app "light meter" which ensured that all pots received the same sunlight, measured in Lux
Ammonium Sulfate [g]	The salt concentration of $(NH_4)_2SO_4$ is a fertilizer for plants. They would support the components of the plant to help them grow (Mosaic Crop Nutrition, 2018).	In order to ensure all the plants received the same amount of fertilizer, the water-salt mixture was poured from one bucket containing 36g of Ammonium Sulfate allowing equal spread and then mixed. It was later poured to every pot equally.

Apparatus

Table 4: Materials needed for the experiment

Materials			
Equipment	**Amount**		**Amount**
Bucket 5L	1x	Pots 10cm diameter	36
Forceps	1x	Trays for pots	7
Scissors	1x	Nails 7cm	60x
Digital Scale [± 0.01g]	1x	Wires with crocodile endings	60x
Digital Multimeter (200m)	1x	Power Supply (Frederiksen-DC) [± 0.1V]	1x
Ruler [±5mm]	1x	Power Supply (MCP-m10-sp303e-DC) [± 0.1V]	1x
Transparent bowl	1x	Beaker 50ml [± 0.1ml]	1x
Paper towel	1x	Spatula	1x
Oven	1x	Permanent Marker	1x
Baking paper	1x	Table Spoon 15ml	1x
Substances			
	Amount		**Amount**
Ammonium Sulfate Pure (Chempur)	30g	Tap Water	18.85L
Universal Soil (COMPO BIO)	5kg	Mung Beans (PNOS)	300x

Figure 2.1 Apparatus set-up

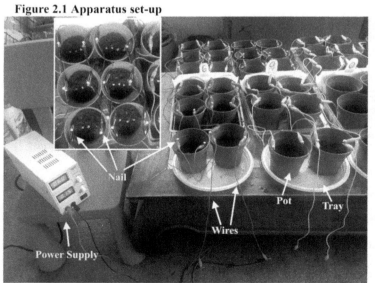

Figure 2.2 Apparatus circuit

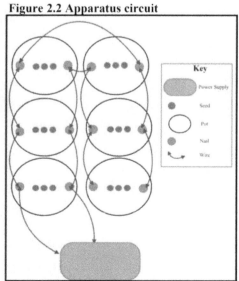

Creating the apparatus

1) 6 pots were arranged on a tray. The tray would collect excess water. 130 grams of soil was added to each pot to fill in 2/3 of the pot. The soil was weight on a digital scale to ensure that all pots have the same amount of soil. Draw a straight horizontal line with a marker on the inside of all the pots where the soil reaches. This would be the starting measuring point. It is advised not to measure from the soil, as the soil could condense with water over time (Measuring Plant Growth, n.d.).
2) 2 nails were pressed inside the soil, to act as conducting electro rod, 8cm apart from each other in each pot.
3) The nail would be 4cm deep inside the soil.
4) Wires with alligator clips were clipped into each nail of the consecutive pot to form a parallel circuit as seen in Figure 2.2. Every nail would have two wires connecting into it, except the last two nails of the last pot. Position the wires so they do not cover the top of the seed. The parallel circuit will allow electricity to continue to flow through the rest of the pots in case one pot fails to conduct the electricity.
5) The first pot would have wires connected from its nail, into the power supply, to allow electricity to flow into the closed circuit. Make sure the power supply is switched off when preparing the apparatus to not get shocked.
6) 3 pregerminated seeds are placed in each pot (with the radicle faced downwards), on top of the soil, arranged in a straight line parallel to the nails. 2cm apart from the next seed and the rode. The arrangement would allow electricity to pass through the seeds when the current is passing from one nail to the other.
7) Sprinkle extra soil in each pot to cover the seeds about 1 cm from the top.
8) With a permanent marker, the apparatus is labeled with the voltage number which would be used on the group.
9) The apparatus is repeated 5 more times for each group
- For control groups of 0V, the apparatus is the same, except that all wires, nails, and power supply were excluded, as electricity is not passing through that group.

Method and Procedure

Preliminary experiment

With a digital multimeter, it was possible to check if there was electricity flowing through the soil, by plugging in 2 test probes in the soil. A test was made to check the conductivity of the soil, by preparing the same apparatus but manipulating the soil. Current was not detected in the soil when it was dry or after being watered. By creating a water-salt mixture and then watering the soil, currents were detected. Water-salt mixture successfully increases conductivity through the soil due to the salts ions, that are charged particles with high mobility to carry charge (Brubaker, 2018). However, the salt water does not allow osmosis through the plant's tissues, causing dehydration in plants and harming them (Lacoma, 2018). An alternative solution with salt was ammonium sulfate $(NH_4)_2SO_4$ which is a salt fertilizer for plants. Furthermore, adding ammonium sulfate to the soil, conducted electricity. It was decided that 1 gram of fertilizer would be added to every pot at the beginning of the experiment, to allow conductivity throughout 3 weeks experiment.

Sample description

For each treatment, 18 seeds were used. 3 seeds placed in each pot, to allow electricity to flow through the plants. With 5 groups exposed to the treatment and 1 control groups, a total of 108 seeds and 36 pots were used to give highly reliable and viable data.

Pre-germinating the seeds

1) The whole package of mung beans (300 seeds) have been dispersed and evenly spread in a bowl.
2) 750ml of warm tap water is added to the bowl and is placed in a window sill to obtain sunlight.
3) Allow the soaked seeds to imbibe in water for 8 hours until a 0.5cm radicle is seen breaking through the seed coating.
4) Select 108 seeds in which the radicle is the same in length and with forceps, carefully place the seed in a towel to dry.
5) Gently add each pre-germinated seed in the soil, to the designated group's apparatus, with forceps, making sure that the radicle does not break.

Preparing the Salt fertilizer

1) With a spatula, measure and weight 36 grams of ammonium sulfate.
2) Place the ammonium sulfate in a bucket and add 3.6 liters of room temperature tap water.
3) Mix the solution thoroughly with a tablespoon until all the ammonium sulfate has dissolved.

Conducting the experiment

1) On day 1 of the experiment, using a measuring beaker, add 100ml of salt fertilizer solution to the soil of each groups pot.
2) Starting from day 2, at 20:00 when the sun is down, the wires of groups would be connected to a power source and set at a specifically designated voltage (3V, 6V, 9V, 12V, or 16V) for 1 hour to allow the current to flow through. The experiment was limited to only 2 power sources, so only 2 groups at random could be exposed to the electricity at the same time and then the power supply is moved to the next groups.
3) After all the 5 groups were exposed to the experiment (0V is not exposed to the electricity), using a ruler, measure the height of any growing plants starting from the horizontal line inside the pot, into the top of the main plant stem. Record the data in a sheet for every seed. Remember that every pot has 3 seeds, so label the seeds from left to right as seed A, B, and C for each pot. Calculate the mean height of each treatment group after every measurement, the groups standard deviation on that day, and the amount of plants recorded.
4) Repeat step 2 and 3 every second day until day 20 of the experiment. Water the plants every day from day 2, with 20ml of tap water in each pot measured with a beaker. Evenly water the pots, from the beaker, in a circular motion to spread out equally. Water the plants at 8:00, to allow the plants to take in water without the excess loss in evaporation and to allow the plants to take in fluid throughout the day to cope with the heat (Rhoades, n.d.).
5) In order to check if electricity is passing through each pot, turn on the power source with the set voltage, and with a multimeter, plug in 2 of its test probes in and view if the current is displayed.

Measuring the Mass

1) On day 20, for each group, carefully with spatula and forceps, remove all the growing plants from the pots making sure the roots do not break and place it in a bowl. Make sure to count how many plants were taken from the group's apparatus. If the roots do break in the process, take the broken root and place it in the same bowl with the plants. Omit the plants which died.
2) Delicately wash off the excess soil stuck in each of the plant's roots in a bowl of water.
3) Place the plants on a paper towel, and gently pat them with another towel to dry out the water and surface moisture.

Wet Mass:

1) Immediately after drying the plants, individually place the plants on a digital scale and record the weight of each plant. Calculate the mean weight and the standard deviation of each experimental group.

Dry Mass:

1) After recording the wet mass, spread out the plants in an oven rack, with a baking paper placed on the bottom. All the plants of the experiment were added in the oven at the same time, separating the groups by adding them to different racks.
2) Leave the plants inside a preheated oven at 35° for 8 hours, enough until the water evaporates from the plants and can be seen visually that it is dried.
3) Cautiously remove the racks from the oven. Using forceps, add the dried plants on a digital scale, and record the weight of each individual plant of each group. Calculate the mean weight and the standard deviation of each experimental group. And then calculate the difference between wet mass and dry mass.

The experiment durations were 20 days to allow sufficient time for the plants to grow and monitor the growth of plants after being exposed to the treatment of voltages. The experiment was conducted during summer (05/06/18 until 25/06/18).

The mean height would be line graphed throughout the experiment and 2 bar graphs would be made for the wet mass and dry mass measured. The results are continuous variable as they result from measuring.

A statistical one-way ANOVA test would be conducted for the mean height on day 20 to check the variance. Additionally, Pearson Correlation would show if there is a significant correlation between the mean height and the voltage. For the mean dry and wet mass, Pearson correlation would also be used to measure the strength associated with the voltage and the mass.

The One-way ANOVA test was calculated using http://www.danielsoper.com/statcalc software website. The Pearson Correlation Coefficient and P-value were calculated using https://www.socscistatistics.com/tests/pearson/Default2.aspx software website.

Ethical, Environmental and Safety issues

As the experiment was made under electrical appliance and live exposed electrodes, safety concerns were a priority. Plastic insulating gloves must always be worn to protect against the electricity. People were always warned not to approach the experiment, as they were live wires, although 16V will not harm a human being. The experiment was conducted in cleanly and safely manner. When using the equipment, be fragile not to damage the glass or power supplies. Use an oven glove when using an oven and remember to turn it off after the experiment. Make sure the power supply is off when setting up the apparatus as the wires are being touched, or when watering the plant, preventing electrical accidents from spilling into you. Make sure all the wires to the power supply is fully insulated and not damaged. Slowly set the power supply to adjust the voltage, as increasing it too fast puts a risk of accidentally setting it to a too high voltage. After the experiment, the equipment was cleaned for future purposes. The organic soil and plant were thrown in a designated organic waste, to make sure the fertilized soil would not get into rivers as it could cause eutrophication to the environment. No animals were harmed during the experiment.

Results and Analysis

The qualitative observation was recorded at the end of the experiment to see how electrical voltages affected the conditions of the plants throughout the 3 weeks experiment.

Table 5: General conditions of plants on the final day

Qualitative Observations	
Treatment	
0V	The plants are all bright in green color, with narrow leaves. Several leaves were curved inwards and crinkled
3V	The plants are all bright in green color, with broader leaves than group 0. Several leaves were crinkled similar to group 0V.
6V	The plants contained yellowing color on the leaves. The leaves were thinner and narrower, but the vitality was still seen.
9V	The stem of the plants was much smaller than other groups. The majority of the leaves were wrinkled. There were signs of black color from the bottom of the stem.
12V	With a lot of dead plants, the stems of the surviving plants were very narrow and dull in black color. The plants were scorched. The majority of the leaves were bent inwards, the blades were not open, but still green. The vitality is this group was low, as slowly the plants started to bend down.
16V	With only 4 surviving plants, the plants were scorched entirely black, with small hints of green color. They were very dehydrated, and the stems were narrowest from all the groups. No leaves opened up in this group. After plunging the plants out of the soil, no roots were seen.

Figure 3: Picture of the experiment on day 20

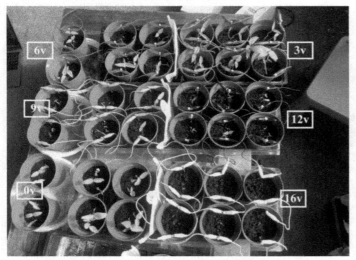

As seen from Figure 3, it was clear that samples treated under high voltage, grew less in height, and a smaller amount of seeds emerged from the ground. As voltage deceased, more of the plants are grown and seen, with the best results in 3V and the 0V (control group), and worst in 16V.

Table 6: Mean height of plant throughout the experiment [± 5mm]

	Treatment groups					
	0V	**3V**	**6V**	**9V**	**12V**	**16V**
Day 2	0.0	0.0	0.0	0.0	0.0	0.0
Day 4	15.0	15.8	13.1	12.0	10.0	0.0
Day 6	31.3	30.0	20.0	18.2	12.7	10.0
Day 8	37.5	33.3	23.8	20.0	13.3	10.0
Day 10	49.4	43.8	28.0	21.4	14.3	10.0
Day 12	60.0	48.8	35.3	24.7	15.7	10.0
Day 14	65.0	55.9	40.6	26.0	18.6	14.0
Day 16	69.4	68.1	48.2	30.0	22.5	15.0
Day 18	72.8	71.3	56.0	39.1	24.5	15.0
Day 20	75.0	74.4	59.3	40.0	28.9	15.0

Graph 1: Mean height of plants throughout the experiment graphed [± 5mm]

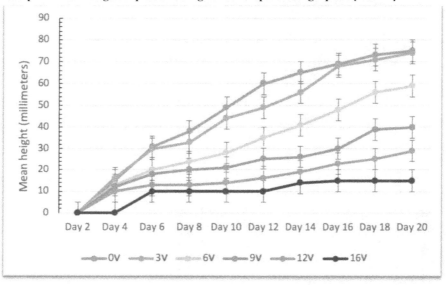

Mean height of plants throughout the experiment

As seen in Graph 1, the graph plotted the mean height of each groups throughout the days of the experiment. An ±5mm error bar was given for each result because of possible eye measurement uncertainties with a ruler. It could be observed that 0V (control group) had the highest mean height throughout the experiment until day 20, where 3V grew to a very similar size. Since day 4, all the samples were already smaller on average compared to control group, except 3V as it had 0.8mm more than the control group. After day 4, all the plants maintained growing gradually, however they also maintained a sparse mean height difference from each other. The lowest mean height was always 16V sample with lowest vitality seen, and then samples with lower voltages, increased average height respectfully and the greenness of the plant. Nonetheless, the control group cultivated the highest mean growth. On the day 20, all groups had significantly different results, as supported by the error bars which do not overlap, except the control group and 3V group which had overlapping error bars making their results are not significantly different. The 3V treatment resulted in highest mean height of 74.4mm compared to other treatments.

Table 7: Collection of mean height of plants on day 20 with standard deviation [± 5mm]

16V	A	B	C
Pot 1	0.0	0.0	0.0
Pot 2	0.0	20.0	0.0
Pot 3	0.0	dead	0.0
Pot 4	0.0	0.0	0.0
Pot 5	10.0	10.0	0.0
Pot 6	0.0	20.0	0.0
5.0	60.0	15.0	4.0

0V	A	B	C
Pot 1	60.0	70.0	60.0
Pot 2	60.0	70.0	60.0
Pot 3	110.0	80.0	80.0
Pot 4	60.0	70.0	90.0
Pot 5	100.0	100.0	30.0
Pot 6	110.0	60.0	80.0
20.3	1350.0	75.0	18.0

3V	A	B	C
Pot 1	60.0	30.0	60.0
Pot 2	140.0	80.0	40.0
Pot 3	80.0	40.0	0.0
Pot 4	70.0	70.0	130.0
Pot 5	90.0	80.0	dead
Pot 6	40.0	90.0	90.0
29.8	1190.0	74.4	16.0

6V	A	B	C
Pot 1	0.0	70.0	80.0
Pot 2	80.0	80.0	dead
Pot 3	70.0	40.0	60.0
Pot 4	dead	30.0	70.0
Pot 5	60.0	30.0	60.0
Pot 6	50.0	40.0	70.0
16.9	890.0	59.3	15.0

9V	A	B	C
Pot 1	dead	30.0	dead
Pot 2	dead	50.0	dead
Pot 3	20	10.0	dead
Pot 4	dead	10.0	50
Pot 5	60.0	dead	0.0
Pot 6	0.0	0.0	90.0
26.0	320.0	40.0	8.0

12V	A	B	C
Pot 1	daed	30.0	30.0
Pot 2	20.0	dead	20.0
Pot 3	dead	10.0	dead
Pot 4	dead	30.0	0.0
Pot 5	0.0	50.0	30.0
Pot 6	dead	40.0	30.0
11.0	260.0	28.9	9.0

Key	
Sum	Total Height
Mean	Average Height
Count	Plants out of the Soil
	Standard Deviation

Example of calculation:

Average height (for 3V) = $\dfrac{\text{Total height}}{\text{Number of plants}}$

$= \dfrac{60+140+80+70+90+40+90+80+70+40+80+30+60+40+130+90}{16} = 74.4\text{mm}$

Mean height during day 20

3V group with the mean height of 74.4mm was very similar and close to the control group with 75mm. However, the standard deviation was much sparser with 29.8 compared to the control group with 20.3. By increasing the voltage to up to 6V, the mean height went down to 59.3mm, but the standard deviation was lower to 16.9. Increasing the voltage to 9V the mean height gradually decreasing, but the standard deviation increased up to 26.0. However, in 12V and 16V, the standard deviation was low with 11.0 and 5.0 respectfully as the number of plants which survived to day 20 was low. The mean height of 12V treatment was 28.9mm, and the lowest mean height was 15.0mm in the 16V treatment.

Statistical Tests for height on day 20

Null Hypothesis (H₀) - Voltages have no significant effect on the growth of the *Vigna radiata* plants
Alternative Hypothesis (H₁) - Voltages have a significant effect on the growth of *Vigna radiata* plants

Analysis of Variance- One-Way ANOVA test

Table 8: Summary of height results

Results during day 20							
Treatments							
	0V	3V	6V	9V	12V	16V	Total
N	18.0	16.0	15.0	8.0	9.0	4.0	70.0
Σx	135.0	119.0	89.0	32.0	26.0	6.0	407.0
Mean	7.5	7.4	5.9	4.0	2.9	1.5	29.3
Std. Dev.	2.0	3.0	1.7	2.6	1.1	0.5	10.9
Σx²	18225.0	14161.0	7921.0	1024.0	676.0	36.0	42043.0

Table 9: Results for One-Way ANOVA test measuring the height

	ss	df	ms	f- value	p-value
Between-Treatments	271.386	5	54.277	11.541	0.00000006
Within-Treatmens	300.997	64	4.703		
Total	52.382	69			

A one-way ANOVA test was conducted to compare the effect of voltages on growth (height). There was a significant effect of the number of voltages on the average height of plants at the p<.05 level for the three conditions [F(5,64) = 11.54, p=0.00000006].

These results suggest that high levels of voltage affect plant height. The p-value of this test is significantly smaller than 0.05, and thus we reject the null hypothesis, and for this test, we accept the alternative hypothesis that voltages have a significant effect on the growth of *V. radiata* seeds.

Table 10: Data for wet and dry mass of the groups

	0V	3V	6V	9V	12V	16V
Wet Mass total [+/- 10mg]	6480.0	6250.0	4930.0	1870.0	2030.0	720.0
Number of Plants	18.0	16.0	15.0	8.0	9.0	4.0
Wet Mass average [+/- 10mg]	360.0	390.6	328.7	233.8	225.6	180.0
Std. Dev.	36.5	58.8	50.4	64.2	6.8	7.1
Dry Mass Total [+/- 10mg]	550.0	480.0	350.0	160.0	120.0	n/a
Number of Plants	18.0	16.0	15.0	8.0	9.0	4.0
Dry Mass average [+/- 10mg]	27.8	29.4	23.5	20.0	13.3	n/a
Std. Dev.	8.5	8.7	9.4	10.0	4.7	n/a
Percentage decrease	92.2%	92.6%	92.7%	91.5%	94.2%	n/a

Mass of the plants

Even though the control group had the highest mean height, it did not have the highest mass, as seen in Graph 2 and Graph 3. The control group had mean wet mass of 360.0mg, relatively less to 3V samples with 390.6mg. This could be explained as the plants in 3V samples had vast and wide leaves, which was not measured when measuring the height. The samples with higher voltage had gradually decreasing wet mass, with the lowest wet mass of 180.0mg in the 16V group. When looking at the results from average dry mass per plant, the control group remained less than the 3V samples. Higher voltage samples had relatively lower dry mass just like the wet mass. Unfortunately, the weight scale could not weight the dry mass of samples from the 16V group, as the weight was smaller than 10mg. Additionally, the results were converted from grams to milligrams. Finally, by calculating the percentage change from wet mass to dry mass, the water content of the plants could have been observed. 12V samples had the most significant percentage decrease in weight of 94.20%; this could explain that it intake the most water proportionally. 9V samples had the lowest percentage decrease of 91.50%.

The error bar for the mean wet mass and mean dry mass was made with 1 standard deviation of each sample, taken from table 10. In Graph 3, the mean dry mass has overlapping error bars from samples because the weight scale was from 10mg and could not give more precise results. In both measurements, the mean wet mass and dry mass did not have significantly different results between the groups as the error bars overlapped each other in every group.

Graph 2: Mean Wet Mass   **Graph 3: Mean Dry Mass**

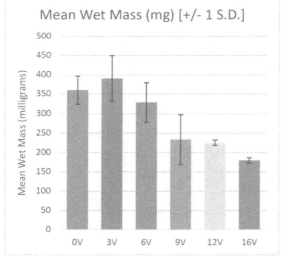

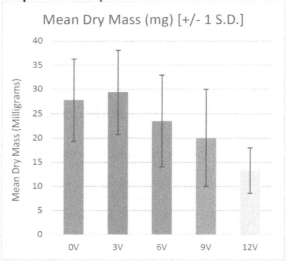

Graph 4: Seeds out of the ground throughout the experiment

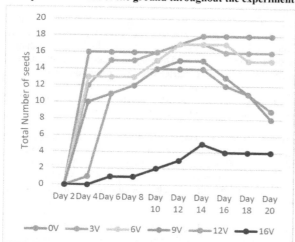

Graph 5: Seeds out of the ground on day 20

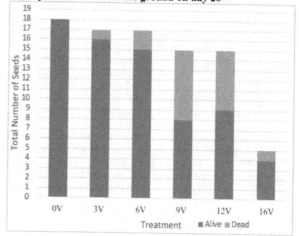

Number of seeds successfully emerging from the ground

As an additional information which arose from the experiment, when measuring the height, some plants started to die and could not be measured anymore until day 20. The plants which were no longer able to record the height were recorded into this graph. Some seeds also did not emerge from the ground, leaving a gap in the data and stunting the growth. In Graph 4, it displays the total number of seeds which successfully emerged from the group out of 18 seeds. If a point decreased, it means that some plants died. It could be seen, that during day 16, plants started to die from samples in 3V, 9V, 12V, and 16V. Out of 18 seeds, all the 18 seeds from the control groups emerged from the ground. 3V and 6V had 1 seed which did not emerge from the ground. The 9V and 12V samples had 15 out of 18 emerged. 16V had the least number of seeds emerging from the ground with only 5 seeds. This could have been the cause of the high voltage under the ground, which did not allow the seeds to continue growing as they died under the soil. At the end of the experiment 9V treatment had the most deaths with 7 deaths out of 15, and then 12V had 6 deaths out of 15 as seen in Graph 5. 16V had only 1 death, but with only 5 seeds successfully growing out of the ground.

Statistical Tests: Pearson Correlation Coefficient

Table 11: Results for mean wet mass, dry mass, and height collected on day 20

Voltages (V)	Mean Wet Mass [±10mg]	Mean Dry Mass [±10mg]	Mean Height [±5mm]
0	360.0	27.8	75.0
3	390.6	29.4	74.7
6	328.7	23.5	59.3
9	233.8	20.0	40.0
12	225.6	13.3	28.9
16	180.0	n/a	15.0

The software website to generate the Pearson R-value plots the X and Y values into its software. X value would be the Voltage, and the Y value would be the Wet Mass, Dry Mass or Mean Height.

Table 12: Pearson Correlation Coefficient Results for day 20

	R Value	P Value
Mean Height	-0.985	0.000336
Mean Wet Mass	-0.939	0.005468
Mean Dry Mass	-0.944	0.015774

R-value results which are near 1 show positive correlation, while results near -1 show negative correlation. The R-Value for all 3 dependent variables are very near -1 with -0.985, -0.939, and -0.944 which explains the independent variable of voltages and the dependent variables have a strong negative correlation with a negative trend line. This means that when there is high voltage, the outcome will be low value of height, wet mass, or dry mass values. All the results of P-value were lower than 0.05 as seen on Table 12 for all 3 dependent variables making the results significant. Furthermore, we reject the null hypothesis that voltages have no significant effect on the growth of the *V. radiata* plants and accept the alternative hypothesis that voltages have a significant effect on the growth of *V. radiata* seeds.

Evaluation of weaknesses, strengths, and suggested improvement

The experiment was very time consuming, as all 108 plants had to be measured individually. However, this gave a range for reliable data. As the uncertainty was only ±5mm, the uncertainty does not compromise the validity of the data in the height measured throughout the experiment. The error bars did not cross suggesting that the results are significantly different. However, when measuring the dry mass and wet mass, the 1 S.D. error bars did overlap in every group, suggesting that the results are not significantly different. The apparatus successfully allowed electricity to go through the ground. The power supply was easy to use, and changing the voltage was precise and straightforward. Measuring several variables in this experiment, such as height and mass, allowed for reasonable and judgmental comparisons between the variables when growing under electrical charges to judge the growth. Using pregerminated seeds ensure that they are in a vital state.

Controlling the controlled variable was essential to keep the results of the experiment reliable so only the independent variable would affect the plants. Throughout the experiment all controlled variables remained the same in all treatment groups. Plants were only considered dead if they fell horizontally to the soil. Deaths were not included in the mean height calculations. As many plants were grown, the shadows from faster-growing plants could have covered the smaller plants, affecting their sunlight intake. Some plants started to bend and directly touched the live electrocuted rods. This could have been avoided if the plants were placed further from each other in bigger pots.

Nail rods started to rust causing rust impurities on the soil which were not considered. The impurities could have toxified the soil affecting the plants' growth. To bypass this problem, non-rusting rods could have been used. The position of seeds was not considered as well. The center seed could have received less voltage, as it was covered with 2 other seeds. To avoid the different positioning in the pot, every pot should have one seed.

When measuring the dry mass and wet mass, the scale was measuring from 0.01g. As the weight of the mass was low, this low-value scale did not give accurate weight of the masses of the plant, making the error bars overlap. The precision of the instrument was not accurate enough to measure the dry mass of 16V samples. For future references, a more precise scale weighing lighter items could be used. When measuring the plant, the size and length of the leaves were not considered, as it was only measured until the top stem. Some plants had huge leaves that were not measured.

As an extension to the experiment, an experiment could be conducted with different treatments from 0V to 5V to narrow the scope of the investigation. The 3V group did not vary much in height with the control group but had higher mass and bigger leaves. Thus, experimenting around 3V could observe positive effects on increasing the plant growth. Furthermore, graphing the rate of growth might also indicate how the voltages affect the growth of the plants.

Conclusion

The experiment supports the hypothesis that voltage helps boost the growth of *V. radiata* up to an optimum level (3V). It is visible in the dry mass and wet mass of the samples, but not the height. 3V sample's mean average height was 74.0mm, compared to 75.0mm in the control group on day 20, however, the wet mass was higher with 391mg in 3V and 360mg in the control group. This is further supported with higher average dry mass of 29.8mg compared to 27.8mg. This shows how much more mass the 3V samples had, boosting their growth. In 3V treatment, where plants' mass increased, the electrical charge stimulated plant cell metabolism which increased ion transport. The increase in transport inside the cells increases the nutritional uptake for the plants (What-when-how, n.d.). This had a positive effect on the mass of the plants.

Over the 3V treatments, the plant's height and mass started to decrease with a significant negative correlation with increasing voltage. Even though the 3V samples had a higher mass, one plant did not grow out of the ground, and it occurred one death, compared with the control group which had no deaths. Every sample under electrical treatment occurred in some deaths and plants unsuccessfully emerging from the ground. The higher the voltage, the fewer plants emerged from the ground. 16V treatments had only 5 out of 18 plants grown, with only 15.9mm in mean height and very scorched black color due to the high voltage burning the plants. Under high voltage, an electric field is a kind of stress, which causes deformation inside grains (Hanafy, Mohamed and El-hady, 2006). The tension on the seeds' biological properties could kill them and damage their tissues which did not allow them to emerge from the ground. 12V samples contained the most water, with 94.20% difference between the wet mass and dry mass. Electricity affects and increases the water uptake of plants (What-when-how, n.d.). This gives more mass made by water uptake. Furthermore, there were factors which occurred unexpectedly throughout the experiment, such as rod rust contaminating the soil.

Experiments made on electrical stimulation on plant growth by Kim and Chun (2017) concluded that the presence of an electrical current is better than a lack of it, but up to 32 uA. It could be seen in an experiment by Novak and Bjeliš (n.d.) where up to 1.5V, beans grew faster than with 0V, but the treatment with higher voltage had a negative impact. Their results match the results from this experiment: plants grow faster up to a certain electrical stimulation. The experiment had two statistical tests with a Pearson correlation coefficient test and one-way ANOVA test that concluded that we should reject the null hypothesis and accept alternative hypothesis with a 95% degree of confidence that voltages have a significant effect on the growth of *V. radiata* seeds on both of the test. Pearson correlation test concluded that all 3 variables (height, dry mass, and wet mass) have a significant negative correlation with voltage. This shows that an increase in voltage, decreases; height, dry mass, and wet mass of the plant.

References

*All the images in the exploration are photographs taken by me.

Akamatsu, F., Kobayashi, A., Hirata, K., Harada, K., Okazawa, A., Hayashi, J. and Takeishi, H. (2013). *Effects of elevated pressure on rate of photosynthesis during plant growth.* [online] Ncbi.nlm.nih.gov. Available at: https://www.ncbi.nlm.nih.gov/pubmed/23994480 [Accessed 5 Sep. 2018].

Brubaker, J. (2018). *Conductivity Vs. Concentration.* [online] Sciencing.com. Available at: https://sciencing.com/conductivity-vs-concentration-6603418.html [Accessed 5 Sep. 2018].

Collins dictionary (n.d.). *Electroculture definition and meaning | Collins English Dictionary.* [online] Collinsdictionary.com. Available at: https://www.collinsdictionary.com/dictionary/english/electroculture [Accessed 4 Sep. 2018].

Hanafy, M., Mohamed, H. and El-Hady, E. (2006). *EFFECT OF LOW FREQUENCY ELECTRIC FIELD ON GROWTH CHARACTERISTICS AND PROTEIN MOLECULAR STRUCTURE OF WHEAT PLANT.* BUCHAREST: Publishing House of the Romanian Academy, p.254.

Jeanty, J. (n.d.). *The Effects of Electricity on Plant Life | eHow.* [online] eHow. Available at: https://www.ehow.com/list_7531004_effects-electricity-plant-life.html [Accessed 4 Sep. 2018].

Kim, B. and Chun, K. (2017). *Electrical Stimulation and Effects on Plants Growth in Hydroponics.* [ebook] Gyeongbuk: Medwell Journals. Available at: http://docsdrive.com/pdfs/medwelljournals/jeasci/2017/4396-4399.pdf [Accessed 5 Sep. 2018].

Lacoma, T. (2018). *What Happens When You Put Saltwater on Plants?* [online] Sciencing.com. Available at: https://sciencing.com/happens-put-saltwater-plants-6587256.html [Accessed 5 Sep. 2018].

Measuring Plant Growth. (n.d.). [Blog] *sciencebuddies.* Available at: https://www.sciencebuddies.org/science-fair-projects/references/measuring-plant-growth [Accessed 5 Sep. 2018].

Mosaic Crop Nutrition. (2018). *Ammonium Sulfate.* [online] Available at: https://www.cropnutrition.com/ammonium-sulfate [Accessed 5 Sep. 2018].

Novak, V. and Bjelis, M. (n.d.). *Effect of electricity on growth and development of several plant species.* [ebook] Zagreb: V Gymnasium. Available at: http://www.hbd-sbc.hr/wordpress/wp-content/uploads/2015/05/CROATIA_Abstract_Novak-and-Bjelis.pdf [Accessed 5 Sep. 2018].

Nytimes.com. (1985). *ELECTRICITY MAY PLAY ROLE IN PLANT GROWTH.* [online] Available at: https://www.nytimes.com/1985/04/09/science/electricity-may-play-role-in-plant-growth.html [Accessed 4 Sep. 2018].

Polygon. (n.d.). *How Humidity Affects the Growth of Plants.* [online] Available at: https://www.polygongroup.com/en-US/blog/how-humidity-affects-the-growth-of-plants/ [Accessed 14 Sep. 2018].

Rhoades, H. (n.d.). *Best Time To Water Plants – When Should I Water My Vegetable Garden?* [online] Gardening Know How. Available at: https://www.gardeningknowhow.com/edible/vegetables/vgen/water-plants-vegetable-garden.htm [Accessed 5 Sep. 2018].

What-when-how (n.d.). *The Effects of Electroculture on Plants - Electro-Horticulture - page 94.* [online] Available at: http://what-when-how.com/Tutorial/topic-1317f/Electro-Horticulture-108.html [Accessed 5 Sep. 2018].

Wheaton, F. (1968). *Effects of various electrical fields on seed germination.* 3521. [online] Retrospective Theses and Dissertations., p.4. Available at: https://lib.dr.iastate.edu/rtd/3521 [Accessed 5 Sep. 2018].

3. CHANGING LIGHT INTENSITIES AND THE EFFECT ON THE PHOTOSYNTHETIC RATE MEASURED USING A COLORIMETER IN ABSORBANCE UNITS (AU) USING ALGAE BALLS IN SEA LETTUCE

Author: Anusha Banerjee
Moderated Mark: 20/24

Research Question: How does changing light intensities in terms of distance [300 mm, 600 mm, 900 mm, 1200 mm, 1500 mm, 1800 mm and 2100 mm (± 0.5 mm)] affect the photosynthetic rate measured using a colorimeter in absorbance units (AU) using algae balls in sea lettuce (*Ulva intestinalis*)?

INTRODUCTION:

This experiment aims to explore the effect of changing light intensity on the photosynthetic rate. A freshwater species of green algae, *Ulva intestinalis*, will be used as it has the capacity to rapidly increase in biomass forming green tides in sea water rich in nutrients (Cummins et al. 2004). To trap the algae, alginate balls will be created with the help of a sodium alginate solution. This is done by the process of immobilization, where sodium alginate solution is mixed with calcium chloride. By immobilizing the algae, they become easier to work with as the trapped algae is kept in one place without losing it and are well preserved for longer. This forms a cross-link between the alginate molecules and sets, making a semi-solid substance that contains the algae trapped inside. After a couple of minutes, the solution hardens, forming strong balls. This method is beneficial as in a short span of time, a large quantity of beads are produced. I decided to use a range of 7 distances; 300 mm, 600 mm, 900 mm, 1200 mm, 1500 mm, 1800 mm and 2100 mm (± 0.5 mm) after research. The algae will be exposed to the light for a duration of 3600 s (+/- 0.0001).

The rate of photosynthesis will be obtained using absorbance units with the help of calorimeter and hydrogen carbonate indicator which evaluates the solution with regards to carbon dioxide levels, thus providing me with tons of qualitative and quantitative data.

BACKGROUND:

Photosynthesis is "the process by which green plants transform light energy into chemical energy. During photosynthesis in green plants, light energy from sunlight is captured and used to convert water, carbon dioxide, and minerals into oxygen and energy-rich organic compounds" (Bassham and Lambers, 2019). This is chemically written as:

$$6CO_2 + 12H_2O + Light\ Energy \rightarrow C_6H_{12}O_6 + 6O_2 + 6H_2O\ \text{(Newton, 2019)}.$$

In photosynthesis, "light energy is used to oxidize water (H2O), producing oxygen gas (O2), hydrogen ions (H+), and electrons. Most of the removed electrons and hydrogen ions ultimately are transferred to carbon dioxide (CO2), which is reduced to organic products" (Bassham and Lambers, 2019). This is a '*light dependent*' reaction since it uses light energy to facilitate the oxidation of water. When carbon dioxide is converted into glucose, this particular reaction is known as '*light independent*' reaction, which involves 'fixing' inorganic CO2 into an organic form of sugar which is later stored in regions known as 'sinks' as a complex sugar, starch (Biology LibreTexts, 2019)

A factor affecting photosynthesis is light intensity, as it unswervingly affects light-dependent reactions in photosynthesis (Brilliant Biology Student, 2019). Light energy is known as "the measure of the wavelength-weighted power emitted by a light source" (Maximumyield.com, 2019). It is collected by a green pigment known as chlorophyll. This pigment is found in the leaves of plants, specifically stored in the chloroplasts. Chlorophyll absorbs light in the form of solar energy and converts it to light energy with the assistance of light harvesting complexes present. The positive correlation is evident as light intensity increases, the light dependent reaction rate increases too. Thus, the rate of photosynthesis increases too, proportionately, increasing the yield of ATP. However, after a while, even though the luminescence may increase, the photosynthetic rate may be limited because of other limiting factors such as temperature. This causes the rate to slow down and form a plateau. An exorbitant increase in light energy may damage the pigment, causing the rate to eventually drop (Rsc.org, 2019).

I chose to investigate this topic as gardening and landscape architecture fascinates me immensely. I enjoy planting and tending to plants, it gives me a sense of responsibility as I strive to keep them alive. To do this successfully, I needed a better understanding of how I can maximize growth and photosynthesis in all my plants. This research will help me conclude the best environment for plants, aiding me in a substantially successful cultivation in my home garden. I selected the algae *Ulva intestinalis* specifically as I read about it in an article I found online. I learnt that hefty quantity of sea lettuce usually pinpoint high levels of pollution (Edc.uri.edu, 2019). When these sea lettuces decompose, ammonia and sulfide is released, the oxygen level in the water decreases, causing a loss of biodiversity. However, this species of algae can be used in wastewater treatment as it has the capacity to remove toxic chromium from wastewater (Anon, 2020). This species can also be used as herbal treatment and as a pharmaceutical ingredient which is beneficial in the prevention of chronic diseases. Hence, I decided to find the ideal light intensity to optimize photosynthesis in this alga.

Around the year 2050, the effects of the several issues regarding the growth in the world's population from its current level of 7.2 billion in 2019 to a predicted total of 9.7 billion will need to be dramatically addressed. This will automatically cause an increase in the want for a different type of diet that involves a higher quality diet which will require greater resources to make sure it is supplied. In order to supply and fulfil the diet of the anticipated 35% increase in the world's population, we need to make some dramatic improvements. To do so, biologists may need to research on new ways in which the efficiency of photosynthesis is increased, especially for those plants used in food. To achieve this, we need to understand the agents that influence the photosynthetic rate.

HYPOTHESIS:
As photosynthesis includes light dependent reactions, light intensity will influence the rate of photosynthesis. As light intensity is highest, where the distance from the plant and light source is least (300 mm), it is hypothesized that the photosynthetic rate will be highest. This will occur as an increased number of photons of appropriate wavelength will strike and be absorbed by chlorophyll (Brennan, 2019). This speeds up the reaction, leading to more photosynthesis. As this occurs, plant produces a greater volume of oxygen as well, whereas the amount of carbon dioxide decreases as it gets used up. Therefore, I hypothesize that the hydrogen carbonate

indicator will turn from color red to color purple when the light intensity is the highest (300 mm) as the CO_2 is used up and its concentration decreases. On the other hand, the light intensity is the least where the distance from the plant and light source is greatest (2100 mm), it is hypothesized that the rate of photosynthesis will be the lowest. Since there is lower light intensity, the number of available photons of appropriate wavelength is lower. Thus, the process of converting light energy into chemical energy will be much slower, slowing down the rate of photosynthesis. Thus, the plant produces lower volumes of oxygen, whereas the amount of carbon dioxide does not decrease majorly as it is not used up. Therefore, I hypothesize that the hydrogen carbonate indicator color may stay the same, red, or change into a yellow-ish orange.

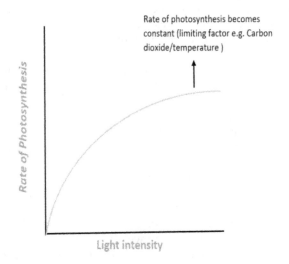

Figure 1.0: The effect of light intensity on the rate of photosynthesis

VARIABLES

Dependent variable: *Absorbance of light (Au) using colorimeter.*
The value of the rate of photosynthesis depends on the different light intensities. The absorbance was used to estimate the photosynthetic rate using a sodium hydrogen carbonate indicator, giving information on the amount of CO_2 absorbed. The more CO_2 absorbed, the greater the rate.

Independent variable: *Light intensity*
Various light intensities were used to show gradual change and trend in the photosynthesis rates. These ranges include 7 distances, 300 mm, 600 mm, 900 mm, 1200 mm, 1500 mm, 1800 mm and 2100 mm ($\pm$ 0.5 mm). By altering the gaps from the light source, a lamp, to the alginate balls, the light intensity is varied. These distances were modified after looking at numerous different experiments online and were pre-tested to ensure a trend is clearly visible when using these distances.

Control variables:
1. *Volume of sodium hydrogen carbonate indicator*
    3ml of the sodium hydrogen carbonate indicator was used for each trial, using a 5mL pipette (+/- 0.01) for increased reliability. This was done as a variation in the value of the indicator may cause a variation in the absorption too, affecting the reliability of the results.
2. *Amount of algae balls in each trial*
    The same number of algae balls, 20, were used for each trial. If the number of algae cells increase, so does the rate for photosynthesis. This affects the results as the absorption will be undertaken by more cells, decreasing the volume of carbon dioxide.
3. *Possible external light sources*
    To limit any other light source, the trials were conducted in a dark room with all windows covered by curtains to block all other light sources. If another external light source was

involved, this would tamper the results as the data would no longer be the result of the lamp's light intensity only.

4. *Temperature of the external environment*
Temperature, just like light intensity, is another limiting factor in photosynthesis. An increase or decrease in the temperature during photosynthesis has a direct correlational result, increasing or decreasing the rate too. To eliminate this confounding variable, the dark room was temperature controlled by setting the air conditioner at 25° Celsius which is the optimum temperature for the sea lettuce to grow.

5. *Time of subjection to light*
The algae will be exposed to the light for a duration of 3600 seconds (+/- 0.0001s). This is ensured by starting a stopwatch timer right after placing the trial on the desired distance from the light source. Time of subjection is maintained as it affects the results. An increase in the amount of time gives the algae cells prolonged time to absorb, possibly resulting in more photosynthesis taking place, interfering with the accuracy of the results.

6. *Phosphate buffer*
A buffer of pH 8 is maintained, to control the optimum pH for the main photosynthetic enzyme Rubisco. This ensures that the enzyme can work at optimum conditions and is not denatured due to changes in pH.

Uncontrolled variable:

1. *Concentration of algae within each bead*
Although the number of beads taken are consistent, when mixing with sodium alginate solution and creating the algae balls it is not possible to control the concentration of algae within each ball. However, to eliminate all possible lumps, it was rotated at the same speed evenly using a magnetic stirrer.

APPARATUS:
- *Ulva intestinalis* (green algae)
- 40 Watts light bulb placed in a lamp
- 3 ml of sodium hydrogen carbonate, for every 5 trials
- 4 ml of sodium alginate solution for every trial
- 7 McCartney bottles for each of the 5 trials, 14 ml each
- 70 ml Calcium chloride solution
- 100 ml Phosphate buffer solution
- 1 gram sodium bicarbonate
- 100 cm ruler (±1.0 mm)
- Beaker (500 ml)
- 1.5 liter distilled water
- Pipette, 5 ml (± 0.015 ml)
- Cuvette
- 10 ml Syringe (± 0.5 ml)
- Sieve
- Stopwatch (± 0.1 second)
- Clamp
- Colorimeter

Method Development:
To develop my method, I carried out extensive research in order to gain insights on the ways in which this experiment can be correctly conducted. During my research, I discovered the site 'www.saps.org.uk', which has abundant resources on experiments regarding Photosynthesis.

Initially, I planned on using neutral density filters. Since the algae balls would be wrapped in these filters, it would allow transmittance of only certain wavelengths of light. An example of this is a 0.3ND filter will allow 50% of the light to transmit through the filter. However, these neutral density filters were unfortunately not available in my school lab, and thus could not be used.

Instead, I decided to use the age-old method of varying distances from a light source in order to get different light intensities. The original protocol of my methodology is taken from the website mentioned above. However, I have tried to reproduce the experiment by modifying the distances at which the algae balls are placed. My aim was to validate and test the existing theory on effect of light intensity on photosynthesis.

Although, the methodology available online relied on visual estimation of color in order to establish a pH value, but to increase accuracy, I decided to use a pH meter and the 550nm filter in the colorimeter to estimate absorbance and measure pH both accurately and reliably.

The independent variable, light intensity, was manipulated by changing the distance of the light source, the lamp, from the bottles containing the algae. This was done using a 100 cm steel ruler, by placing the bottles at the varying distances which were measured and marked in mm (± 0.5 mm).

METHOD:
    A. The green algae, *Ulva intestinalis* was placed into a beaker with 250 ml of water to settle for 30 minutes. This was done to concentrate the active algal culture. Using a pipette, the pale green suspension of water was extracted, while the concentrated algae remained undisturbed.
    B. Sodium alginate solution is created by mixing water and alginate powder together with the help of a magnetic stirrer.
    C. In order to completely dissolve, the solution is left overnight.
    D. 4 ml (± 0.015 mL) of the sodium alginate was added to the concentrated algal mixture and gently mixed, avoiding any production of bubbles.
    E. A syringe is clamped to a stand, positioned right above a beaker of 70 ml calcium chloride solution.
    F. The algal solution is poured slowly, through the barrel of a 5mL syringe (8.66 mL diameter), allowing it to pass and drip into the beaker placed vertically below it, forming beads. As the drops fell, the beaker was gently swirled.
    G. 100 ml of a phosphate buffer solution at pH 8 is added to ensure enzyme activity is optimum.
    H. The mixture was left to set for 10 minutes, which forms bead-like shapes of the algae.
    I. The algae balls were then strained out using a sieve, while being washed with distilled water to increase their shelf life.

J. 20 perfectly round undamaged algae balls were meticulously positioned in each McCartney bottle.
K. 0.5 grams of sodium bicarbonate was mixed with 500 ml of water to create a 0.1% solution of sodium bicarbonate.
L. The sodium bicarbonate solution was added to each bottle using a pipette.
M. Using a pipette, 3 ml of sodium hydrogen carbonate indicator was added to each bottle.
N. The distances of 300 mm, 600 mm, 900 mm, 1200 mm, 1500 mm, 1800 mm and 2100 mm were marked from the light source (lamp) using a 100 cm ruler, where each bottle was respectively positioned.
O. A heat filter (a beaker of water) was placed between the light source and the algal balls in order to ensure that the variation in temperature is controlled.
P. Simultaneously, the lamp was switched on along with a stopwatch for 3600s (+/- 0.0001).
Q. Using a sensor extension cable, the colorimeter was connected to the data collection system.
R. The colorimeter was calibrated so it reads an absorbance of 0.0 for the solvent, which is distilled water.
S. The cuvette is held at the lid and wiped using a non-abrasive cleaning tissue.
T. The cuvette is placed in the colorimeter and the lid is closed.
U. The calibrate button is pressed.
V. Once calibration is complete, the cuvette is removed from the colorimeter.
W. The color and absorbance value (Au) of the sodium hydrogen carbonate indicator was observed and noted using the colorimeter at 550 nm.
X. Readings were taken 5 times for each distance.
Y. After the completion of 3600 seconds, the lamp is switched off and the hydrogen carbonate indicator is removed from the immobilized algae.
Z. When removing the indicator, it is vital that no extra carbon dioxide is accidently added. Thus, pipettes are used, that have air extracted from them prior to using them.

Risk Assessment:
Safety
In order to ensure maximum safety, a few protocols are followed when using the chemicals in the experiment. Firstly, a lab coat was worn at all time to avoid spillages and accidents on skin. Moreover, alginates are known to be irritants to the eyes and respiratory track. To ensure it is safely used, lab safety goggles were used, and no direct inhalation of the powder or solution took place. Calcium chloride is said to be extremely reactive if swallowed, leading to burns, vomiting, etc. To avoid this, it was stored in a secure area away from incompatible materials and moisture.

Environmental considerations
Although hydrogen carbonate solution is not harmful and has no known toxicity, it was disposed off by draining it out in a foul water drain and traces were cleaned using water. In order to dispose Calcium chloride solution, state laws were followed to safely discard the waste chemical. Similarly, the phosphate buffer solution was transferred into a chemical waste container and disposed of according to local regulations. The algae used was a type of invasive species, thus, care was taken to prevent it from going into water bodies and instead was placed in optimum conditions in the laboratory aquarium and used for other photosynthesis experiments.

Ethical considerations
No animals were used during experimentation hence there were no ethical considerations.

RESULTS:
Qualitative Data:

Table 1 shows the change in color and average pH obtained after 3600 seconds.

	Distance from light source (mm) (± 0.5 mm)						
	300	600	900	1200	1500	1800	2100
Initial colour	Red	Red	Red	Red	Red	Red	Red
Colour change	Purple	Red (no change)	Pink	Dark orange	Orange	Light orange	Yellow
Average pH	8.7	8.4	8.3	8.2	8.0	7.8	7.6

Figure 2.0: Color chart used for interpretaing pH values

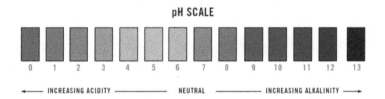

The table above shows the change in color from the beginning of the experiment to after the completion of 3600 seconds. It can be observed that the greater the distance from the light source, the lighter the color of the indicator and lower the pH value.
These results can be explained by the amount of CO_2 absorbed and the CO_2 concentration left inside the bottles. The bottle that was the furthest away from the lamp turned into the lightest color (yellow), with the lowest pH value (7.6) compared to other bottles. The change in color from dark (red) to light (yellow) shows increasing CO_2 in the indicator. This shows that photosynthesis is not taking place since CO_2 is released via the process of respiration, instead of being absorbed via photosynthesis. In contrast, the bottle that was the closest to the lamp turned into the darkest color (magenta), with the highest pH value (8.7) compared to other bottles. The change in color from red to magenta shows decreasing levels of CO_2 in the indicator. As photosynthesis occurs, it removes the CO_2 from the indicator. This supports the hypothesis. When the algae balls are closer to the light source, the rate of photosynthesis is higher as the absorption of CO_2 increases.

Raw Data: Refer to Appendix
Processed Data:

Table 2 shows the light intensity, average absorbance value of indicator and standard deviation of all 5 trials.

	Distance from light source (m) ($\pm$ 0.0005 m).						
	0.300	0.600	0.900	1.200	1.500	1.800	2.100
Relative light intensity ($1/D^2$) (lx)	11.11	2.77	1.23	0.69	0.44	0.31	0.23
Average absorbance value of indicator (Au)	0.75	0.55	0.43	0.40	0.35	0.24	0.17
Standard deviation	0.007	0.004	0.004	0.004	0.013	0.004	0.008

The standard deviation calculated above is using an online calculator. The calculation of standard deviation helps identify how dispersed a set of data is, from its average. Since the values are very small, this means they are all close to the average, establishing reliability of the results and the 5 trials. The calculation for the sample of 2100 mm is shown next to this information.

The light intensity was however calculated using a graphic display calculator.
The formula to calculate light intensity is:
$$\frac{1}{D^2}$$
The formula above was applied, however, the units of distance (mm) had to be converted into SI units (m) in order to correctly use this formula.

For example,
300 mm = 0.300 m

Thus,

$$\frac{1}{D^2} = \frac{1}{0.3^2} = \frac{1}{0.09} = 11.1111 \ldots$$

Standard Deviation, s: **0.0089442719099992**

Count, N: 5
Sum, Σx: 0.87
Mean, $\bar{x}$: 0.174
Variance, s^2: 8.0E-5

Steps

$$s = \sqrt{\frac{1}{N-1} \sum_{i=1}^{N} (x_i - \overline{x})^2,}$$

$$s^2 = \frac{\Sigma(x_i - \bar{x})^2}{N - 1}$$

$$= \frac{(0.17 - 0.174)^2 + \ldots + (0.19 - 0.174)^2}{5 - 1}$$

$$= \frac{0.00032}{4}$$

$$= 8.0E\text{-}5$$

$$s = \sqrt{8.0E\text{-}5}$$

$$= 0.0089442719099992$$

The rate of the reaction calculation is shown in the table below.

Table 3 shows the rate of the reaction

	Distance from light source (m) ($\pm$ 0.0005 m).						
	300	600	900	1200	1500	1800	2100
Absorbance value (Au)	0.75	0.55	0.43	0.40	0.35	0.24	0.17
Calculation	$\dfrac{0.75}{3600}$	$\dfrac{0.55}{3600}$	$\dfrac{0.43}{3600}$	$\dfrac{0.40}{3600}$	$\dfrac{0.35}{3600}$	$\dfrac{0.24}{3600}$	$\dfrac{0.17}{3600}$
Rate of Reaction (Au/s)	0.00021	0.00015	0.00012	0.00011	0.00009	0.00006	0.00004

The graph displaying the rate of reaction is attached in the appendix.

Below is a line graph that represents the data from Figure 3.0. This graph illustrates the relationship between the values of the average absorbance values of the indicator and the relative light intensity. The error bars are obtained using the calculation from standard deviation.

Figure 3.0

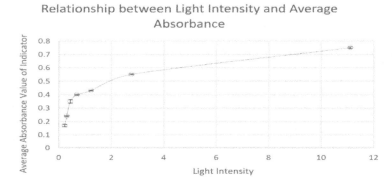

What can be concluded from the graph above is that the lower the light intensity (0.23), implying the alginate beads were further away from the light source (1200 mm), the lower the average absorbance value seen (0.17). Vice versa, the higher the light intensity (11.11), the closer the alginate beads were from the light source (300 mm) and thus a higher average absorbance value is observed (0.75). The error bars are very small and hardly visible on the graph, suggesting that the error calculated in the results was very low, making the data accurate. The standard deviation was also quite low, suggesting that all values obtained are centered around the mean, again increasing the validity and reliability of the results.

The Anova statistical test was conducted in order to investigate if the treatments were significantly different from each other. It does so by grouping the means between different groups and comparing them.

Null Hypothesis: There will be no significant correlation observed between the distance of algae balls from the light source and the absorbance value of the indicator.

Alternative Hypothesis: There will be a significant correlation observed, positive, between the distance of algae balls from the light source and the absorbance value of the indicator.

Table 4 shows the Results of the Anova Test

Source	Sum of squares SS	Degrees of freedom V	Mean square MS	F statistic	p-value
Treatment	1.3377	6	0.2229	4,777.5000	1.1102e16
Error	0.0016	35	0.0000		
Total	1.3393	41			

From this test, the p-value corresponding to the F-statistic is lower than 0.05. This suggests that only one or more of the treatments are significantly different. To further analyze and gain a clearer understanding, the Tukey HSD test was done, in order to pinpoint which pair of treatments were significantly different from one another.

Table 5 shows the results of the Tukey Test

Treatments Pair	Tukey HSD Q Statistic	Tukey HSD p-value	Tukey HSD interference
A vs B	71.7137	0.0010053	**p<0.01
A vs C	113.5467	0.0010053	**p<0.01
A vs D	124.3038	0.0010053	**p<0.01
A vs E	142.2322	0.0010053	**p<0.01
A vs F	182.8700	0.0010053	**p<0.01
A vs G	206.1769	0.0010053	**p<0.01
B vs C	41.8330	0.0010053	**p<0.01
B vs D	52.5901	0.0010053	**p<0.01
B vs E	70.5185	0.0010053	**p<0.01
B vs F	111.1563	0.0010053	**p<0.01
B vs G	134.4632	0.0010053	**p<0.01
C vs D	10.7571	0.0010053	**p<0.01
C vs E	28.6855	0.0010053	**p<0.01
C vs F	69.3233	0.0010053	**p<0.01
C vs G	92.6302	0.0010053	**p<0.01
D vs E	17.9284	0.0010053	**p<0.01
D vs F	58.5662	0.0010053	**p<0.01
D vs G	81.8732	0.0010053	**p<0.01
E vs F	40.6378	0.0010053	**p<0.01
E vs G	63.9447	0.0010053	**p<0.01
F vs G	23.3070	0.0010053	**p<0.01

LEGEND	
Alphabet	**Distance from the lamp (cm)**
A	300
B	600
C	900
D	1200
E	1500
F	1800
G	2100

As seen above, the treatments are significantly different from each other. This explains that there is a strong chance that the numerical change that can be observed is related to another change. Thus, I can accept my alternative hypothesis and conclude that there is a significant correlation distance of the algae balls and the absorbance values obtained.

Correlation Coefficient

The Pearson's correlation coefficient statistical test was also conducted in order to measure the strength and direction of the linear relationship between the two variables. It was done using an online calculator on the site: https://www.socscistatistics.com. This calculation is shown in the appendix.

The correlation calculation has shown that there is a correlation between the absorbance value of the indicator and the light intensity, as seen by the graph above as well. A strong positive correlation between the two variables is noticeable, which means the X-values go with the Y-values. This correlation suggests that when light intensity increases, the absorbance rate of the indicator increases too. However, as the extent from the light source increases, the absorbance rate of the indicator decreases as light intensity decreases too. Moreover, the critical value at 1 degrees of freedom is 0.997, which is greater than the R value found, 0.8721. Thus, the null hypothesis can be rejected and the alternative hypothesis can be accepted, that is, the light intensity does have an effect of rate of photosynthesis. Where the light intensity was the highest, so was the rate of photosynthesis, as the absorbance value of the indicator was the highest. The greater distance that the algae balls were from the lamp, the lower the absorbance value, and thus,the rate of photosynthesis was lower.

CONCLUSION

In conclusion, my research supported my hypothesis, proving that varying light intensities do affect the rate of photosynthesis in the algae species *Ulva intestinalis*. The closer this alga is to light, as in, the greater light intensity it is exposed to, the greater the rate of photosynthesis taken place, as at a distance of 300 mm, the average absorbance value recorded was 0.75 Au, whereas at a distance of 2100 mm, the average absorbance was lower, 0.17 Au, suggesting a lower rate of photosynthesis at a further distance.

The data that supports this is reliable since its standard deviation is low. This shows that all values are centered around the mean, which removes possibilities of anomalies. The data obtained is similar to the scientific theory and supports it, ensuring reliability and accuracy. Moreover, the correlational calculation also shows a strong positive relationship, providing more support for the relationship hypothesized for the two variables. The Anova and Tukey post HOC tests also backed up the hypothesis, as a significant difference was observed, which means a strong chance of correlation is present.

Thus, my hypothesis is supported. This can be explained as light intensity increases, the algae are exposed to more energy which can be converted from light intensity into chemical energy, which is used in photosynthesis. Light is also used in photolysis reaction during photosynthesis, where water molecules are broken down in the presence of light. This is seen through the color change, from red to magenta, indicating a higher pH level. The higher pH shows decrease in acidity, implying that carbon dioxide concentrations were low and oxygen concentrations were higher.

Moreover, research on *Scenedesmus obliquus* by one of the researchers, Barbara Gris, observed the effect of the main parameter that impacts microalgae, which is light. This was assessed by cultivating the above species in a flat plate photobioreactor. It showed a maximum growth rate of

150 μmol photons m(-2) s(-1) and for any value below this, growth was limited due to the factor of light. When increased, effects of photo saturation on the cells were observed (Gris, 2013).

In addition to this, research was done by Constantine Sorokin and Robert W. Krauss on different species of algae, including *Chlorella pyrenoidosa, Chlorella vulgaris, Scenedesmus obliquus and Chlamydomonas reinhardti*. They found increased light intensity increased the rate of photosynthesis, but at high light intensities photosynthesis is inhibited in all the algal species they studied (Sorokin and W. Krauss, n.d.).

As an extension, I could possibly use colored filters such as red, blue, green and violet, wrapped around each McCartney bottle. As each filter is of a different color, the wavelengths reflected will also be different. Using this, the impact of varied wavelengths on photosynthetic rate can be studied as photo-dependent reactions will rely on the wavelength of light available.

EVALUATION

One of the strengths of my experiment is the wide variety of distances used. This provides me with more data and increases the extent to which I can analyze the pattern and rate of photosynthesis. I also used a colorimeter, which is a digital apparatus and thus the chances of human error were minimized, providing accurate data to interpret. 7 light intensities with 5 trials were done and the data collected provided a trend in the effect of light intensity on photosynthetic rate. Not only this, but the qualitative data also matched the quantitative data. As observed, the photosynthetic rate increased and so did the recorded absorbance values of the indicator.

Another feature of this experiment which is a strength is the lack of a limiting factor. Instead of water, a sodium bicarbonate solution was used, ensuring that carbon dioxide was readily available. This eliminates the possibility of carbon dioxide being a limiting factor which could have potentially could have caused inaccurate results had it been limiting since photosynthesis itself would be restricted.

Another limitation is that the color of the solution was determined manually in the qualitative observation. Since color can sometimes be objective, the chances of human error arise by misjudging shades. Hence a pH meter was used to obtain quantitative accurate readings to arrive at a proper conclusion.

Not only this, but since the distances were measured by hand, using a ruler, there could have been the possibility of difference of a few millimeters due to human error, known as parallax error, which occurs due to viewing an object from a different angle,(parallax error – due to viewing it at a different angle), which could have perhaps played a key role in the data variability. Similarly, using pipettes and measuring cylinders also create the possibility of human error. This error can be corrected by increasing the number of readings and finding the average.

Although the algae balls were relatively similar in size, in order to ensure maximum accuracy and efficiency in terms of uniform surface area of the beads, a device known as the Spherificator can be used. This device is handheld and can form spheres of the same shape due to the presence of nozzles, which allow the production of up to 250 beads in a short span of time.

BIBLIOGRAPHY

Saps.org.uk. (2020). *'Algal balls' - Photosynthesis using algae wrapped in jelly balls*. [online] Available at: https://www.saps.org.uk/secondary/teaching-resources/235-student-sheet-23-photosynthesis-using-algae-wrapped-in-jelly-balls [Accessed 20 Feb. 2020].

Socscistatistics.com. (2020). *Social Science Statistics*. [online] Available at: https://www.socscistatistics.com/ [Accessed 20 Feb. 2020].

Biology Articles, Tutorials & Dictionary Online. (2020). *Photolysis Definition and Examples - Biology Online Dictionary*. [online] Available at: https://www.biologyonline.com/dictionary/Photolysis [Accessed 20 Feb. 2020].

Statistics Solutions. (2020). *Table of Critical Values: Pearson Correlation - Statistics Solutions*. [online] Available at: https://www.statisticssolutions.com/table-of-critical-values-pearson-correlation/ [Accessed 20 Feb. 2020].

Online.visual-paradigm.com. (2020). *VP Online - Online Drawing Tool*. [online] Available at: https://online.visual-paradigm.com/app/diagrams/#proj=0&type=LineChart [Accessed 20 Feb. 2020].

Biology LibreTexts. (2020). *Home*. [online] Available at: https://bio.libretexts.org/ [Accessed 20 Feb. 2020].

Brilliant Biology Student. (2020). *Biology Study Guide*. [online] Available at: http://brilliantbiologystudent.weebly.com/ [Accessed 20 Feb. 2020].

Encyclopedia Britannica. (2020). *photosynthesis | Importance, Process, Cycle, Reactions, & Facts*. [online] Available at: https://www.britannica.com/science/photosynthesis [Accessed 20 Feb. 2020].

Maximumyield.com. (2020). *MaximumYield - Cultivating a Healthy Life for You & Your Plants*. [online] Available at: https://www.maximumyield.com/ [Accessed 20 Feb. 2020].

2020, F. (2020). *The Royal Society of Chemistry*. [online] Rsc.org. Available at: https://www.rsc.org/ [Accessed 20 Feb. 2020].

Edc.uri.edu. (2020). *Geospatial Solutions for Environmental Challenges*. [online] Available at: https://www.edc.uri.edu/ [Accessed 20 Feb. 2020].

Cummins, S.P., D. E. Roberts, and K.D. Zimmerman. 2004. Effects of the green macroalga *Enteromorpha intestinalis* on macrobenthic and seagrass assemblages in a shallow coastal estuary. Marine Ecology Progress Series 266: 77-87.

Ncbe.reading.ac.uk. (2020). [online] Available at: http://www.ncbe.reading.ac.uk/PRACTICALS/PDF/Catmilk1.5_UK_eng.pdf [Accessed 2 Mar. 2020].

Anon, (2020). [online] Available at: https://www.researchgate.net/profile/Ahmed_El_Nemr/publication/6447325_Removal_of_toxic_chromium

4. THE EFFECT OF ETHANOL CONCENTRATION ON THE HEART RATE OF DAPHNIA MAGNA

Author: Marco Buttigieg
Moderated Mark: 21/24

The main goal of the experiment was to determine the impact of an increase in ethanol concentration on the heart rate of several *Daphnia magna* specimens. This was done by placing the specimens in solutions of various concentrations of ethanol and measuring the change in heart rate of each specimen.

D. magna are invertebrate crustaceans, more commonly known as water fleas, which range in size from 2-5mm. They can be found in freshwater ecosystems all across the northern hemisphere. *D. magna* are known to feed on microorganisms such as zooplankton and phytoplankton, but also can consume bacteria, as well as fungal spores like yeast. Their bodies are enclosed in a transparent shell called a carapace that is composed of the polysaccharide chitin (Elenbaas, 2013). Unlike most other crustaceans, *D. magna* possess myogenic hearts, which can generate their own action potentials to cause contractions in the cardiac muscles. This makes the cardiovascular system of *D. magna* more similar to that of larger vertebrates like humans than that of other invertebrates, who have neurogenic hearts that rely on signals from the brain (Yamagishi et al., 2000).

When working with live specimens such as *D. magna*, some ethical considerations must be made to ensure the safety of the organisms. A published study on the effects of melatonin and ethanol on the heart rate of *D. magna* utilized an ethanol concentration of 5%. Researchers found no evidence of harm to the organisms following trials, so 5% was deemed a safe concentration of ethanol for the organisms to survive in (Kaas et al., 2009). With the organisms regularly inhabiting a wide variety of bodies of water, they must also endure a variety of conditions such as water temperature and pH. *D. magna* can survive in pH conditions ranging from 6.5-9.5 (Ebert 2005) and temperatures around 21°C (Carolina Biological Supply Company, 2019). It was also critical to use spring water as opposed to distilled water whenever dealing with *D. magna* because it does not contain trace amounts of harmful chemicals such as chlorine which can harm the organisms (Cornell University and Penn State University, 2009).

Ethanol, otherwise known as alcohol, is a common organic psychoactive chemical compound with the chemical formula C_2H_6O. It is typically used as a recreational drug, consumed within beverages such as beers or distilled spirits (Collins and Kirouac, 2013). Alcohol affects the brain by increasing the effects of an inhibitory neurotransmitter called gamma-aminobutyric acid (GABA) (Harris and Lobo, 2008). Neurotransmitters are chemical messengers that transmit signals in the nervous system (Lodish et al., 2000). Ethanol allows GABAergic receptors in synapses to remain open enhancing the effects of GABA (Harris and Lobo, 2008). GABA causes Cl⁻ ions to enter the post-synaptic neuron, hyperpolarizing the neural membrane, thus inhibiting the neuron (Watanabe et al., 2002).

The heart rate of each specimen was recorded before and after immersion in the ethanol solutions by observing the contractions of the organism's heart in a 240fps (slow motion) video. This gave an initial and final heart rate measurement in beats per minute (bpm). Using these two measurements, percent change in heart rate was calculated using the following formula:

$$\% \text{ Change Heart Rate} = [(HR_{Following\ Submersion} - HR_{Resting}) / HR_{Resting}] \times 100\%$$

The results of the calculation were then analyzed in order to determine a conclusion to the initial question.

On a recent family trip to Cuba, I learned that it is legal to consume alcohol at the age of 16. As I drank at the resort we were visiting, I realized that alcohol had a very interesting effect on my body. After a few drinks, I began to feel quite sluggish and had difficulty keeping my balance. Slowly, after I stopped drinking, these effects began to subside. If I drank again, I felt the same way. Later, I learned that alcohol does this because it is a psychoactive depressant, which has a strong effect on my nervous system. Running such an experiment would be quite difficult to control with human subjects, but when I found that *Daphnia magna* were affected by alcohol in a similar way, I thought that allowing these organisms to consume alcohol would be an interesting way to investigate the effects of alcohol on the body.

Experimental Design

Research Question:

What is the effect does increasing ethanol concentration have an effect on the percent change in heart rate of *D. magna* following immersion?

Hypothesis:

If ethanol concentration is increased, then the heart rate of the *D. magna* will decrease because more ethanol will be absorbed into the body of the organism, so a greater inhibitory effect will be observed, thus decreasing the frequency of cardiac nerve signals that cause contractions.

Variables:

Table 1: Ethanol Concentration and Percent Change in Heart Rate Variables

	Variable	Method of Measurement
Independent Variable	Ethanol Concentration (% V/V)	Five solutions were created (0%, 0.50%, 1.00%, 1.50%, 2.00%) by diluting an 80.8% stock ethanol solution with spring water
Dependent Variable	Percent Change in Heart Rate (%)	The heart rate (bpm) of each specimen was measured both before and immediately following immersion in the ethanol solution, then percent change was calculated using the appropriate formula

Table 2: Controlled Variables during Experimentation

Controlled Variable	Significance of Variable	Method of Control
Approximate Size of *D. magna*	Specimens of different sizes would have different metabolic rates, allowing them to better/worse metabolize the ethanol in the solution, thus changing the extent of the drug's effect on heart rate	All specimens selected for experimentation were of approximately 2-5mm in size, indicating that the specimen was matured, giving the most uniform metabolic rates possible for testing
Room where Experiment was Conducted	Room conditions including air temperature, lighting, and humidity can all have an effect on the results of the experiment	All trials were performed in the same room in the same place to ensure uniform conditions for all trials
Temperature of Culture Solution	Temperature could potentially have an impact on the heart rate of the specimens, so changes in solution temperature could affect the change in heart rate	Due to water's high specific heat capacity, small fluctuations in room temperature would not have a significant effect, so the water temperature of solution was left at a temperature around 21°C throughout the experiment; within ethical boundaries
pH of Culture Solution	Changes in pH making the solution overly acidic or basic can harm the specimens if it exceeds safe levels	Spring water has a pH of ~7, thus it was safe for the specimens
Rest Prior to Experimentation	If specimens are given different rest periods prior to trials, some will not have yet reached their resting heart rate while others may have drifted into an overly relaxed state, making for larger differences in results between specimens	Each specimen was given 10 minutes in spring water in order to reach its resting heart rate
Duration of Heart Rate Measurement	Measurements lasting for different amounts of time can cause inconsistencies in heart rate	Each video taken to measure heart rate was 30 seconds long, multiplied by two to find beats per minute
Time Elapsed for Immersion in Ethanol Solution	Varying lengths of time spent in the solution would cause the specimens to absorb different amounts of the drug, which could cause large discrepancies in the change in heart rate	Each specimen would spend 5 minutes the container of solution, then an additional 1 minute in the drop of solution, totaling 6 minutes spent in each ethanol solution
Timing of Heart Rate Measurement Following Immersion	As time progresses after immersion, the heart rate of the specimens will slowly return to normal, so the timing of each measurement needed to be the same	Following immersion, a 1-minute window was given to immobilize the specimen so it could be observed, and the video was started promptly after 1 minute

Camera Used	The camera used allowed for the collection of heart rate data, so different cameras recording at different frame rates and resolutions could result in some heart beats not being detected	All videos for heart rate monitoring were shot at 720p and 240fps by the same iPhone 7 camera
Time Elapsed between Feeding of Specimens	Specimens that have been starved for different amounts of time will have different metabolic rates, directly impacting their heart rates	Specimens were fed once every 24h during trials with 3mL of yeast solution

Materials and Apparatus:

- 5 Mature *D. magna* (Ward's Science)
- 1162.62mL Spring Water*
- 12.38mL 80.8% Ethanol Solution* (Anachemia)
- 6 250mL Beakers (±5mL)
- 2 10mL Graduated Cylinder (±0.01mL)
- Laboratory Film
- 2 Plastic Pipettes
- 7 Petri Dishes
- White Paper
- Lamp
- 10x Tripod Magnifier (Compass)
- iPhone 7
- Vaseline
- Yeast
- Hot Plate
- Glass Stirring Rod
- Stopwatch (±0.01s)
- Thermometer (±0.5°C) (-20.0°C – 140.0°C)

*Assumed pure, no uncertainty

Procedure:

1. 5 mature *D. magna* specimens were taken from the culture with a plastic pipette and placed in a petri dish containing the culture solution
   i) 75mL of spring water was added to an empty petri dish
   ii) The petri dish was left for 2 hours to allow water to reach 21°C (room temperature)
   iii) *D. magna* were transferred from original culture to culture solution.
2. Five 200mL ethanol solutions were prepared (0%, 0.50%, 1.00%, 1.50%, 2.00%)
   i) 200mL of spring water was measured in a 250mL beaker (0%)
   ii) 1.24mL of 80.8% ethanol was measured in a 10mL graduated cylinder and diluted with 198.76mL of spring water in a 250mL beaker (0.50%)
   iii) 2.48mL of 80.8% ethanol was measured in a 10mL graduated cylinder and diluted with 197.52mL of spring water in a 250mL beaker (1.00%)

    iv)   3.71mL of 80.8% ethanol was measured in a 10mL graduated cylinder and diluted with 196.29mL of spring water in a 250mL beaker (1.50%)

    v)   4.95mL of 80.8% ethanol was measured in a 10mL graduated cylinder and diluted with 195.05mL of spring water in a 250mL beaker (2.00%)

    vi)   All 5 solutions were left on the lab table covered with laboratory film until use

3. The heart rate measuring apparatus was prepared

    i)   Vaseline was smeared at the base of a petri dish

    ii)   Petri dish was placed on a sheet of white paper

    iii)   The lamp was positioned beside the petri dish and turned on

    iv)   The tripod magnifier was placed over the centre of the petri dish

    v)   iPhone 7 was placed on top of the tripod magnifier, with the camera focussed at the base of the petri dish

4. Specimens were timed 10 minutes in the spring water solution with the stopwatch

5. ~50mL of 0% ethanol solution was poured into a petri dish

6. At the 10 minute mark, specimens were transferred onto the heart rate measuring apparatus with a plastic pipette and placed under the magnifying lens

7. A 30-second video was taken with the iPhone 7 at 720p and 240fps

8. The specimens were transferred into the 0% ethanol solution with a plastic pipette

9. The specimens were immersed in the 0% ethanol solution for a 5-minute interval

10. At 5 minutes, the specimens were returned to the heart rate measuring apparatus where they remained for an additional 1 minute

11. After the additional 1 minute, another 30-second video was taken by the iPhone 7 at 720p and 240fps

12. The videos were played back and both resting and ethanol influenced heart rate was counted by observing the contractions of the heart on the video and multiplying the number by 2 to achieve the heart rate in beats per minute

13. Steps 5-13 were repeated for each specimen at each ethanol concentration

14. Percent change in heart rate for each specimen at each ethanol concentration was calculated

15. Mean and standard deviation of percent change in heart rate for each ethanol concentration was calculated

16. Mean results were plotted on a graph with a trendline

Analysis

Observations:

Table 3: Raw Concentration and Heart Rate Measurements of Each *D. magna* Specimen

Ethanol Concentration (% V/V)	Volume of Ethanol Stock Solution (mL) (±0.01mL)	Volume of Spring Water (mL) (±0.01mL)	Specimen Number	Resting Heart Beats (/30s) (±0.5 beats)	Ethanol Influenced Heart Beats (/30s) (±0.5 beats)
0	0	200.00	1	84.0	85.0
			2	99.5	93.5
			3	98.0	98.0
			4	90.5	90.0
			5	98.5	99.5
0.50	1.24	198.76	1	85.5	82.5
			2	100.5	95.0
			3	97.0	96.0
			4	89.5	85.5
			5	98.5	93.5
1.00	2.48	197.52	1	85.0	78.0
			2	98.5	91.5
			3	98.0	92.0
			4	90.5	82.0
			5	99.0	88.0
1.50	3.71	196.29	1	82.5	70.0
			2	102.0	87.5
			3	96.5	84.5
			4	92.5	79.5
			5	98.0	85.5

2.00	4.95	195.05	1	86.0	70.5
			2	99.5	82.0
			3	99.0	84.5
			4	91.5	70.0
			5	100.0	81.0

Table 4: Qualitative Observations of *D. magna* Specimens During Experimentation

Ethanol Concentration (% V/V)	Stage of Trial	Observations
0	Before Submersion	Specimens were quite active, swimming frequently in the petri dish
	During Submersion	Specimens were quite active, swimming frequently in the petri dish → Specimen 1 reproduced
	After Submersion	Specimens were quite active, swimming frequently in the petri dish
0.50	Before Submersion	Specimens were quite active, swimming frequently in the petri dish
	During Submersion	Activity of specimens was slightly slowed → Specimen 3 swimming regularly → Specimen 5 reproduced
	After Submersion	Specimens regained all activity
1.00	Before Submersion	Specimens were quite active, swimming frequently in the petri dish
	During Submersion	Specimens noticeably less active, swimming slower → Specimens 3 and 4 reproduced
	After Submersion	Specimens regained all activity
1.50	Before Submersion	Specimens were quite active, swimming frequently in the petri dish
	During Submersion	Specimens lost most activity, swimming very slow
	After Submersion	Specimens had gained back some activity → Specimen 3 regained regular activity
2.00	Before Submersion	Specimens were quite active, swimming frequently in the petri dish
	During Submersion	Specimens stopped all swimming, completely inactive → Specimen 3 moving very slowly
	After Submersion	Activity slowly regained among specimens, slow swimming → Specimen 4 was still not active at this point

Data Processing:

Overview:

The data collected in the experiment was processed by calculating the percent change in heart rate of each specimen at each ethanol concentration. The percent change in heart rate was used because it shows how the specimen's heart rate changes while accounting for the individual differences in heart rate between specimens. The mean percent change in heart rate was calculated for each ethanol concentration was calculated along with its sample standard deviation. The mean helps to standardize results within each set of trials to gain a better representation of the general outcome, and the standard deviation shows how data within ea̲ 0.01mL/4.95mL x 100% = 0.2% ation was used as opposed to popul̲a̲t̲i̲o̲n̲ ̲s̲t̲a̲n̲d̲a̲r̲d̲ ̲d̲e̲v̲i̲a̲t̲i̲o̲n̲ because only e experiment, as opposed to the entire population. The uncertain 5mL/200mL x 100% = 2.5% count for the random errors which could have occurred during the experiment. Finally, the results of the experiment were plotted on a scatter graph comparing ethanol concentration (% V/V) on the x-axis and percent change in heart rate (%) on the y-axis, along w̲i̲t̲h̲ ̲a̲ ̲l̲i̲n̲e̲ ̲o̲f̲ ̲b̲e̲s̲t̲ ̲f̲i̲t̲ ̲t̲o̲ ̲s̲h̲o̲w̲ ̲t̲h̲e̲ ̲g̲e̲ 2.7%/100% x 0.02 = 0.0005

Sample Calculations:

Stock Ethanol Solution Dilution (2.00%):

$$C_1 V_1 = C_2 V_2$$

$$V_1 = \frac{2.00\% \times 200 \text{mL}}{80.8\%}$$

$V_1 = 4.95 \pm 0.01 \text{mL}$ of stock ethanol solution should be added to 195.05mL of spring water to make a 200mL 2.00% V/V solution

Ethanol Concentration Uncertainty (2.00%):

$$C = \frac{V_{Ethanol}}{V_{Solution}} \times 100\%$$

$$C = \frac{4.95 \pm 0.01 \text{mL}}{200 \pm 5 \text{mL}} \times 100\%$$

$$C = \frac{4.95 \text{mL} \pm 0.2\%}{200 \text{mL} \pm 2.5\%} \times 100\%$$

$$C = (0.02 \pm 2.7\%) \times 100\%$$

$$C = (0.02 \pm 0.0005) \times 100\%$$

$$C = 2.00 \pm 0.05\%$$

Beats Per Minute (Specimen 1, 2.00%, Resting):

Heart Rate = Heart Beats Counted x 2

Heart Rate = 86.0 ± 0.5 beats x 2

Heart Rate = 172 ± 1 bpm

Percent Change in Heart Rate (Specimen 1, 2.00%):

$$\% \text{ Change Heart Rate} = \left(\frac{HR_{\text{Following Submersion}} - HR_{\text{Resting}}}{HR_{\text{Resting}}}\right) \times 100\%$$

$$\% \text{ Change Heart Rate} = \left(\frac{141 \pm 1\text{bpm} - 172 \pm 1\text{bpm}}{172 \pm 0.5\text{bpm}}\right) \times 100\%$$

$$\% \text{ Change Heart Rate} = \left(\frac{-31 \pm 2\text{bpm}}{172 \pm 1\text{bpm}}\right) \times 100\%$$

$$\% \text{ Change Heart Rate} = \left(\frac{-31\text{bpm} \pm 6.45\%}{172\text{bpm} \pm 0.58\%}\right) \times 100\%$$

$$\% \text{ Change Heart Rate} = (-0.180 \pm 7.03\%) \times 100\%$$

$$\% \text{ Change Heart Rate} = (-0.180 \pm 0.013) \times 100\%$$

$$\% \textbf{ Change Heart Rate} = \mathbf{-18.0 \pm 1.3}\%$$

Mean Percent Change in Heart Rate (2.00%):

$$\bar{x} = \frac{1}{N}\sum\nolimits_{i=1}^{N} X_i$$

$$\bar{x} = \frac{1}{5}\sum\nolimits_{i=1}^{5} X_i$$

$$\bar{x} = \frac{(-18.0 \pm 1.3\%) + (-17.6 \pm 1.1) + (-14.6 \pm 1.1) + (-23.5 \pm 1.2) + (-19.0 \pm 1.1)}{5}$$

$$\bar{x} = \mathbf{-18.5 \pm 1.2}\%$$

Sample Standard Deviation (Percent Change in Heart Rate, 2.00%):

$$Sx = \sqrt{\frac{1}{N-1}\sum\nolimits_{i=1}^{N}(X_i - \bar{x})^2}$$

$$Sx = \sqrt{\frac{[-18.0 - (-18.5)]^2 + [-17.6 - (-18.5)]^2 + [-14.6 - (-18.5)]^2 + [-23.5 - (-18.5)]^2 + [-19.0 - (-18.5)]^2}{4}}$$

$$\mathbf{Sx = 3.22}\%$$

Table 5: Processed Heart Rate Measurements of Each *D. magna* Specimen

Ethanol Concentration (% V/V)	Specimen Number	Resting Heart Rate (bpm) (±1bpm)	Ethanol Influenced Heart Rate (bpm) (±1bpm)	Percent Change in Heart Rate (%)	Mean Percent Change in Heart Rate (%)	Sample Standard Deviation (%)
0	1	168	170	1.2±1.2	-0.9±0.9	2.95
	2	199	187	-6.0±1.0		
	3	196	196	0		
	4	181	180	-0.6±1.2		

86

	5	197	199	1.0±1.0		
0.50±0.02	1	171	165	-3.5±1.2	-3.9±1.1	1.80
	2	201	190	-5.5±1.0		
	3	194	192	-1.0±1.0		
	4	179	171	-4.5±1.1		
	5	197	187	-5.1±1.0		
1.00±0.03	1	170	156	-8.2±1.2	-8.4±1.1	1.96
	2	197	183	-7.1±1.1		
	3	196	184	-6.1±1.0		
	4	181	164	-9.4±1.2		
	5	198	176	-11.1±1.1		
1.50±0.04	1	165	140	-15.2±1.3	-13.7±1.1	1.13
	2	204	175	-14.2±1.0		
	3	193	169	-12.4±1.1		
	4	185	159	-14.1±1.2		
	5	196	171	-12.8±1.1		
2.00±0.05	1	172	141	-18.0±1.3	-18.5±1.2	3.22
	2	199	164	-17.6±1.1		
	3	198	169	-14.6±1.1		
	4	183	140	-23.5±1.2		
	5	200	162	-19.0±1.1		

Figure 1: Percent Change in Heart Rate of *D. magna* Following Submersion in Different Ethanol Concentrations

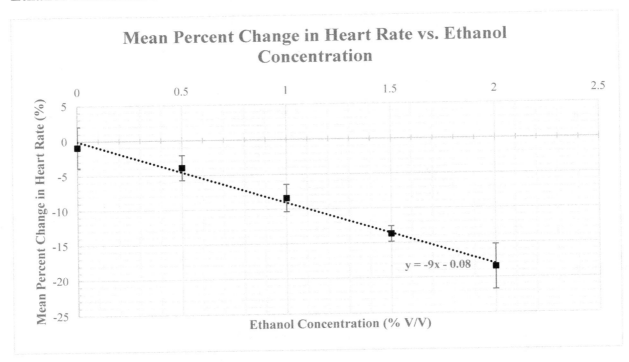

Note: Error bars represent the sample standard deviation

Conclusion and Evaluation

Conclusion:

Once the experiment was concluded and results were analyzed, a strong trend was discovered in the data in the relationship between ethanol concentration (%V/V) and the depressant effect of the drug on the invertebrate species *D. magna*, seen through the percent change in heart rate of the specimens. It is important to note that the higher the absolute value of the percent change is, the more of a depressant effect the drug has on the body of the *D. magna*. As percent change in heart rate approaches zero, the drug has a weaker impact in the organism. At the lowest concentration, 0% ethanol, the solution was pure spring water. There was little to no observed changes in the heart rate of the *D. magna*, with a mean percent change of merely -0.9%. As ethanol concentration increased to 2.00%, percent change in heart rate was greatest, averaging 18.5% among the specimens. As ethanol concentrations were increased within this range, a negative linear correlation was observed between it and the percent change in heart rate of the *D. magna*, with all other data points falling on or very close to the trendline. This means that as ethanol concentration increases, the heart rate of *D. magna* will decrease, supporting the initial hypothesis. Thus, it can also be concluded that as ethanol concentration increases, it has a positive linear relationship with the depressant effect of the drug on *D. magna*.

This was the expected result of the experiment, following the proposed scientific theory. Ethanol has a depressant psychoactive effect on the body by enhancing the effect of the primary inhibitory neurotransmitter, GABA. It does this by allowing GABA sensitive receptors in the synapse to remain open for an extended period, so the neurotransmitter has a stronger effect (Harris and Lobo, 2008). The increased effect of GABA allows more Cl^- ions to enter the postsynaptic neuron, making it significantly harder for it to create and propagate an action potential due to the hyperpolarization (Watanabe et al., 2002). With the inhibitory neurotransmitter having a stronger effect, the heart of the *D. magna* will have a more difficult time sending signals to contract, thus lowering heart rate. As ethanol concentrations increased in the experiment, the amount of ethanol absorbed into the body of the *D. magna* increased accordingly, thus causing the linear increase of the depressant effect of the drug on the specimens. It is important to note here that this trend would not continue indefinitely as ethanol concentration increased because, at a certain point, a lethal dose of ethanol would be administered, killing the *D. magna*. Though there was no experiment in literature completed using the same concentrations of ethanol, a published study investigating this change at a 5% concentration of ethanol also saw a sharp decrease in heart rate following submersion (Kaas et al., 2009), similar to the decreases seen in this experiment.

Evaluation:

In the experiment conducted to determine the relationship between ethanol concentration and the percent change in heart rate of *D. magna* specimens, there were numerous sources of systematic error that should be addressed. The most prominent error was observed during the transfer of the specimens from their spring water resting solution to the ethanol solution, and from the ethanol solution to the heart rate measuring apparatus. The transfer, done with a plastic pipette, was a disturbance in the activity of the specimen. This disturbance is characterized by the increase in heart rate observed after trials at 0% ethanol, pure spring water. Theoretically, the solution was no different from the resting solution, thus no change in heart rate should have been observed. This would have caused less of a decrease in heart rate of each specimen, as heart rate would increase after the final transfer. This error could have been resolved if the ethanol was added directly to the spring water to make the solutions rather than using premade solution, and instead of transferring the specimens to the heart rate measuring apparatus, the solution was drained from the petri dish and the apparatus was moved to the drained dish.

Another systematic error that was observed in the experiment was seen in the reproduction of the *D. magna* observed during trials. *D. magna* are known to reproduce both sexually and asexually when breeding partners are absent (Ebert, 2005). The reproduction of the specimens during trials would have been an additional stress on the organisms, presumed to increase heart rate. This increase in heart rate during trials would have caused the heart rate following submersion to be higher than what it should be when influenced by ethanol. This error could have been removed if, prior to each trial, specimens were examined under a microscope to determine if any developing young were present in the body of the specimen. Trials should have only been commenced if it was certain that the specimens would not reproduce during trials.

A final error in the experiment is the inconsistency between specimens during testing. Each *D. magna* specimen would respond to and metabolize ethanol slightly differently. This was exhibited by the varying responses observed among the specimens during testing with regard to their movement, and the amount of time taken to fully recover after exposure to the ethanol. This would have had opposing effects on the results, with some specimens having a greater change and others having a smaller change, but in such a small sample size, these minute differences could significantly sway the results. The only way to resolve this error would have been to increase the sample size, with a larger sample of perhaps 25 specimens would provide a better representation of the actual trend that exists for the organisms.

An improvement that could have been made to the experiment is to incorporate more ethanol concentrations within and beyond the ranges used in the experiment. Doing this could show at which concentrations ethanol can be lethal in *D. magna*, as well as better representing how the effects of ethanol change as concentration increases. A further reach would be to investigate the effects of other psychoactive drugs on the heart rate of *D. magna*, as it provides good insight as to how these drugs also affect the human heart.

5. THE EFFECT OF E-CIGARETTE USE ON THE HUMAN SALIVARY A-AMYLASE ACTIVITY MEASURED THROUGH THE SPECTROPHOTOMETRIC ABSORBANCE OF MALTOSE

Author: Yara Alzubi
Moderated Mark: 21/24

Research Question: 'What is the Effect of Changing the Vegetable Glycerin Concentration in E-cigarette liquid (50%, 60%, 70%, 80%) on 30 μL of Human Salivary α-amylase Activity as Measured Through the Spectrophotometric Absorbance of Maltose (AU) Post Adding 1.0000 mL of 5% starch solution?'

1. Personal Engagement

This IA was chosen to be explored on the subject of e-cigarettes to raise more awareness on their dangers to human health; especially as I increasingly read about the life-threatening cases related to e-cigarette use that are most frequently happening around the world to people of my age group. This research question was decided to be answered in particular after witnessing first-hand an often overlooked side effect of using e-cigarettes, low taste satiety, on my friend; as my curiosity triggered me to investigate the reason behind this side effect. Which I later learned is a side effect of low enzyme activity due to the saliva's exposure to vapors produced from e-liquids of high vegetable glycerin concentrations.

2. Exploration

2.1 Introduction

Electronic cigarettes – most commonly referred to as 'e-cigarettes' – are battery-powered aerosol-burning devices used to mimic and/or alternate the conventional tobacco cigarettes ("Vaping Devices (Electronic Cigarettes)," 2020). They were first invented by Chinese pharmacist Hon Lik in 2003 (Brueck, 2019); however, their use was popularized across the globe only in the last decade (Rahman et al., 2014). Specifically, the world has witnessed an increase of 34 million e-cigarette users from 2011 to 2018 (Jones, 2019). As for many of its users, it is seen as a healthier alternative to tobacco cigarettes and a potential helper to quit smoking. Consequently, scientists and researchers have taken interest in investigating the chemical and physical make-up of the device – especially as e-cigarette-linked diseases increase each year.

Most research is done on the pulmonary side-effects of the inhaled vapor in e-cigarettes; whereas little of it was concerned with its side-effects on the saliva of the inhaler. Specifically, little to no research has been done on its effect on the salivary α-amylase enzyme activity; although such approach has been taken in numerous investigations on the side-effects of tobacco cigarettes (Nagler et al., 2000; Weiner et al., 2009). Researching the effects of inhaled e-cigarette vapor – vaporized from e-liquids of different concentrations of vegetable glycerin – on the salivary α-amylase enzyme activity is of particular significance considering the activity's association with the human glycemic homeostasis measure (Mandel & Breslin, 2012) and taste satiety (Nakajima, 2016).

2.2 Hypothesis

If the vegetable glycerin concentration in the e-cigarette's e-liquid is increased (50%, 60%, 70%, 80%), then its user's salivary α-amylase enzyme activity measured through the spectrophotometric absorbance of maltose (AU) is expected to decrease. The proposed cause is that vegetable glycerin produces acrolein gas upon vaporization; an aldehyde which causes the α-amylase enzyme in the saliva to undergo carbonylation when the user orally inhales the e-cigarette gas. In turn, the accumulation of carbonylated α-amylase enzymes will inhibit the overall enzyme activity.

2.3 Background knowledge

The electronic cigarette activates when its user's inhalation subsequently creates airflow that triggers the atomizer to vaporize the e-liquid in its cartridge which, eventually, is inhaled by the user (Rahman et al., 2014). Although e-cigarette manufacturers claim that the device's vapor is clear of the 4000 known harmful chemicals in cigarette smoke (Rahman et al., 2014), it still shares common ground with some of the cigarette's infamous chemicals: nicotine (mg/ml), vegetable glycerin (%), and propylene glycol (%) (Devito & Krishnan-Sarin, 2018).

Acrolein (C_3H_4O) is an unsaturated aldehyde ("Acrolein," n.d). It is the product of a heat-induced dehydration reaction of glycerol (Stevens & Maier, 2008); which is 'a clear, colourless, viscous, sweet-tasting liquid belonging to the alcohol family of organic compounds' ("Glycerol," n.d). Commercially, glycerol is known as 'glycerin'; which is slightly less pure than glycerol (Lai et al., 2012). The glycerin found in e-cigarette aerosol is vegetable glycerin (VG), and is used in a percentage ratio to the other base component; propylene glycol (PG) (Phillips et al., 2017) – both of which function as humectants which constitute for 70-80% of the e-liquid (Kosmider et al., 2014). E-cigarette users may choose which percentage ratio they prefer when purchasing their e-liquid – with some opting for percentages of VG that are higher than PG considering VG's higher vapor density, sweet taste it produces, and other reasons. When the atomizer in the e-cigarette's vaporization chamber heats the e-liquid (Cassidy, 2019) to temperatures of 180 °C or greater (Stevens & Maier, 2008), the VG (oil) is then oxidized to acrolein (Ogunwale et al., 2017). Similar to the relative abundance of methylglyoxal, formaldehyde, and acetaldehyde (from the oxidation of PG) in e-cigarette vapor (Bekki et al., 2014), the abundance of acrolein in each e-cigarette vapor is controlled by the concentration of VG in the cigarette's e-liquid.

The α-amylase-containing saliva is secreted into the human oral cavity from the two parotid glands. This saliva is comprised of water, electrolytes, mucus, and enzymes (Tiwari, 2011). Primarily, it functions in maintaining oral hygiene, lubricating and binding of food, solubilizing dry food, and initiating starch digestion (Tiwari, 2011). When exposed to aldehydes (in gaseous state) such as acrolein, a number of the salivary enzymes undergo carbonylation – a form of protein modification where threonine, an α-amylase amino acid side chain (Scannapieco et al., 1989), undergoes metal-catalyzed oxidation; thus forming a carbonyl group in the enzyme structure which results in enzyme inactivation (Wong et al., 2012) (this was investigated using cigarette smoke aldehydes) (Ros, 2017). A distinct salivary protein that is negatively-affected by this exposure is the α-amylase enzyme; which is responsible for catalyzing the hydrolysis of starch (polysaccharide) into maltose (disaccharide) (Bagchi & Nair, 2018) in the human mouth. Since, the accumulation of carbonylated α-amylase enzymes inhibits the overall enzyme activity (Nagler et al., 2000).

A spectrophotometer is 'a device used to measure absorbed light intensity as a function of wavelength.' (Nilapwar et al., 2011, p. 64); whose digital measurement is based off the amount of detected photons absorbed after transmitting through the chosen solution ("Spectrophotometry," 2019). In a UV-visible spectrophotometer, the absorption of the analyte of interest in the solution is detected upon the color change (of the color reagent used) that occurs in its presence. Furthermore, when applied to spectrophotometry, the Beer-Lambert law proposes that the absorbance reading of the analyte of interest is directly proportional to its concentration (Wypych, 2018).

2.4 Variables

Independent Variable: Change in the concentration of VG in e-cigarette e-liquid (%) *Range:* 50%, 60%, 70%, 80%. This variable was changed based on the VG concentration smoked by each of the chosen e-cigarette saliva volunteers; with every set of 10 volunteers smoking a different concentration level than the other set. Furthermore, the concentration smoked by each set was found by viewing it on the label of the e-cigarette e-liquid (commercially known as juice) used by each volunteer. Lastly, this specific range of concentrations was chosen because of its availability; as it was highly difficult to find e-cigarette smokers that smoked VG at concentrations out of this range (making it the normal range smoked).

Dependent Variable: Maltose light absorbance reading (AU) ± 0.01. Values for the light absorbance of maltose in each sample were found by reading them on the spectrophotometer.

Controlled Variables:

1   Subject inclusion criteria. Each unit in the criteria was controlled based on the testimonials of the volunteers that informed their consent to participate (Check Appendix B); which were obtained through the distributed survey (Check Appendix A). As for why each unit was controlled is as follows:
    - For all volunteers: No known disorder/disease; no disorder/disease is a variable of interest in the investigation, 18 and above; legal smoking age in my country (Jordan) ("A Law Gone up in Smoke," 2017), fast; food or drink remnants in the oral cavity may affect the volunteer's normal α-amylase activity.

- Specific to e-cigarette volunteers: Smoking history, e-cigarette voltage and vaporizing temperature, and smoking pattern; considering that difference in any of the above variables will account for a difference in the e-cigarette volunteer's level of exposure to acrolein (higher battery output voltage and vaporizing temperature increases the production of acrolein) (Kosmider et al., 2014); an extensive variance in any of the above variables will likely cause for an inequality among the α-amylase activity results of the e-cigarette smoking volunteers. Moreover, each of the variables' specific range (or exact value in the case of the e-cigarette's voltage and vaporizing temperature) was chosen because it was found to be the prevalent one among the chosen smoking volunteers, restricted e-cigarette use; no type of smoking other than e-cigarettes is a variable of interest in the investigation.

2 Healthy non-smoking saliva samples (0% VG). The nature of the samples (healthy non-smoking) was controlled based on the testimonials of their volunteers.

3 Room temperature in which each saliva sample was collected. Collected all saliva samples outside where the weather in Amman (my city) at the time of collection was recorded at 26°C during the day on October 21st, 2019 ("AccuWeather," n.d). This was to ensure all of the saliva samples went under the same temperature conditions prior to the assay.

4 Mode of saliva collection. All volunteers were instructed to passively drool saliva into vials made of hydrol class clear glass material; to make sure there was no tampering in the type of saliva provided.

5 Vial type. All volunteers were instructed to passively drool saliva into the same vials made of hydrol class clear glass material.

6 Vial size. All volunteers were instructed to passively drool saliva into the same 5.00 mL ± 0.05 vials.

7 Saliva samples' storage refrigeration temperature. The refrigerator was set at 4°C ± 1; the temperature at which the α-amylase in the saliva samples can remain stable for up to 14 days (Barranco et al., 2018).

8 Spectrophotometer's wave length. The light wavelength of the spectrophotometer was calibrated before each test to 660 nm; which is the wavelength at which potassium iodide (color reagent) is absorbed at maximum in the experiment (Cheesbrough, 1999)

9 Room temperature where each saliva assay was executed. In the laboratory, room temperature was set to 25°C ± 1 using the air conditioner; a temperature at which α-amylase can remain stable under for up to 5 days (Moore et al., 1999).

10 Incubation temperature. The water bath was set at a temperature of 37°C ± 1 in preparation for the experiment, and was the only instrument used for incubation all throughout. Incubation was specifically controlled at this temperature in an attempt to imitate the temperature of the environment where α-amylase is normally activated (the human oral cavity), in addition to it being the enzyme's optimum temperature in the mouth.

11 Type of incubator. All saliva assay solutions were placed in the same water bath incubator.

12 Type of color reagent, starch solution, water and buffer. The same solutions of color reagent (1% potassium iodide), 5% starch, and PBS were pipetted from in the assay of every saliva sample.

13 Volumes of saliva, water, starch, PBS, and color reagent. 30.0 μL ± 0.9 of saliva, 21.0 ± 0.5 distilled water, 1.0000 ml ± 0.0006 starch, 1.0000 ml ± 0.0006 buffer, and 2.00 ml ± 0.01 of color reagent were pipetted into every saliva assay test tube.

14 Blank composition. All saliva assay solutions were read on the spectrophotometer against a blank composed of only distilled water.

15 Type of water. Only distilled water was used throughout the whole experiment.

Uncontrolled Variables: Duration between saliva sample collection and α-amylase assay. This was monitored by collecting all the samples and executing their assays in the same 24-hour period; which was to ensure all of the saliva samples went under similar environmental conditions prior to the assay.

2.5 Apparatus
1- (10) healthy non-smoking saliva samples.
2- E-cigarette user saliva samples; (10) exposed to 50% VG, (10) to 60%, (10) to 70%, and (10) to 80%.
3- (50) 5.00 mL ± 0.05 vials.

4- Spectrophotometer (AU) ± 0.01.
5- 50.0 mL 5% starch solution.
6- (1) 50.0 mL ± 0.5 graduated cylinder.
7- (1) 50.0 mL boiling tube.
8- Water bath (1.0°C to 99.9°C) ± 0.1.
9- (1) micropipette (5.0 to 50.0 μL) ± 0.9.
10- (50) 50.0 mL ± 0.5 test tubes.
11- (1) micropipette (100.0 to 1000.0 μL) ± 0.6 μL.
12- 50.0 mL 1X phosphate buffered saline (PBS) pH 7.0 (buffer).
13- 100.0 mL 1% Potassium iodide solution (color reagent).
14- 1053.5 mL Distilled water.
15- (1) 5.00 mL ± 0.01 volumetric pipette.
16- Labels.
17- Black marker.
18- Stopwatch ± 0.01 s .
19- Digital thermometer (-50.00 to 150.00°C) ± 0.01.
20- Sodium hypochlorite (5% available chlorine) solution.
21- Envirocide disinfectant.

2.6 Subject Inclusion Criteria

For all saliva sample volunteers, it is essential that none of them has any known disorder/disease and are 18 and above. Additionally, they are all required to fast for at least 8 hours prior to sample collection. As for the e-cigarette user saliva sample volunteers, the inclusion criteria that is specific to them is as follows:
1- They have been e-cigarette users for between 6 to 12 months.
2- During their e-cigarette use period, they have not attempted any other type of smoking i.e. tobacco cigarettes.
3- Their e-cigarettes are set to a battery output voltage of 3.7 volts, and a vaporizing temperature of 250 °C.
4- They have a similar smoking pattern, where they each smoke between 4-6 days in a week; using up 1 mL of e-cigarette e-liquid each day (which is equivalent to approximately 100 puffs). The 1 mL was measured by the number of times the user re-fills his/her e-cigarette cartridge; as their cartridges had a maximum capacity of 1 mL.

2.7 Methodology

1- Give one labeled (with its volunteer's identifying code) vial to each volunteer and instruct him/her to passively drool saliva into it from the opening.
2- Refrigerate each sample at 4°C ± 1 after collection.
3- Calibrate the spectrophotometer to 660 nm, and set the lab's air conditioner temperature to 25°C ± 1 and the water bath to 37° C ± 1.
4- Spray the laboratory countertop with envirocide disinfectant then wipe.
5- Measure 50.0 mL of the prepared starch solution using the graduated cylinder and pour it into the "starch" labeled boiling tube, then incubate the tube in the water bath and time 5 minutes on the stopwatch.
6- Pipet 30.0 μL ± 0.9 of each saliva sample using the 50.0 μL ± 0.9 measuring micropipette into its assigned test tube.
7- Pipet 1.0000 mL ± 0.0006 of the PBS buffer solution using the 1000.0 μL ± 0.6 measuring micropipette into each test tube then mix by inversion.
8- Place all tubes in the same water bath (from step 5) then time 5 minutes on the stopwatch.
9- Check if the starch solution's temperature has reached 37.00°C ± 0.01 using the digital thermometer. If that temperature has been reached, proceed to step 10.
10- While continuing to incubate, pipet 1.0000 mL ± 0.0006 mL of the warmed starch solution using the 1000.0 μl ± 0.6 measuring micropipette into each test tube then time 7 minutes on the stopwatch.
11- Check if the temperature of each test tube has reached 37.00°C ± 0.01 using the digital thermometer – whilst rinsing the thermometer with tap water between each tube check. If that temperature has been reached, proceed to step 12.

12- Remove the tubes from the water bath then pipet 2.00 mL ± 0.01 of potassium iodide solution (color reagent) using the volumetric pipette into each one.

13- Pour 21.0 mL ± 0.5 of distilled water using the graduated cylinder into each test tube then mix by inversion. Therefore, each test tube now contains 30.0 μL ± 0.9 of saliva, 1.0000 mL ± 0.0006 PBS, 1.0000 mL ± 0.0006 starch, 2.00 mL ± 0.01 potassium iodide, and 21.0 mL ± 0.5 of distilled water.

14- Record the detected color change of each saliva assay solution.

15- In one cuvette, pipette 3.50 mL ± 0.01 of distilled water using the volumetric pipette to be used as the blank calibrating each test in step 12.

16- Use the spectrophotometer to test the light absorbance of each saliva assay solution by pipetting 3.50 mL ± 0.01 from the chosen solution using the volumetric pipette into another cuvette and record the reading.

17- When finished, wash out all the saliva assay solutions using sodium hypochlorite solution.

18- Spray the laboratory countertop with envirocide disinfectant then wipe.

2.8 Justification

Saliva was collected from each volunteer using the passive drool method because of its efficiency. Furthermore, this procedure was adapted from the α-amylase method proposed by the biochemist Wendell T. Caraway in his paper titled 'A Stable Starch Substrate for the Determination of Amylase in Serum and Other Body Fluids' (Caraway, 1959) considering its sound reasoning and applicability (in comparison to other α-amylase assay methods found in literature such as that of Nelson Somogyi's).

For similar to Caraway: after the procedure's analyte of interest (saliva) was pipetted into its assigned test tube; a pH 7.0 buffer (PBS) was added to adjust and maintain the assay solution's pH at a value that is optimum for the activity of the human salivary α-amylase enzyme (Rudeekulthamrong & Kaulpiboon, 2012). Afterwards, starch was added to the solution to allow the α-amylase in the saliva to hydrolyze it into maltose. Potassium iodide was then added as a color reagent to detect the disappearance of starch and change the color of the solution accordingly, followed by a recording of the visually-detected color. The same assay method was carried out on each of the 50 saliva samples (10 non-smoking control samples (0% VG), 10 exposed to 50% VG, 10 to 60%, 10 to 70%, and 10 to 80%) through several incubation periods to speed up the chemical reactions.

Finally, a spectrophotometer was used to measure the color intensity of the solution after the hydrolysis reaction.

2.9 Risk Assessment

Safety Issues: First, a laboratory protective lab coat, goggles and nitrile gloves were worn during the experiment in the laboratory at all times. Second, laboratory countertops were sprayed with envirocide disinfectant then wiped before and after the experiment. Third, all saliva assay solutions were washed out using sodium hypochlorite (5% available chlorine) solution prior to their disposal through the sink.

Ethical Issues: A comprehensive informed consent form was given out to every saliva sample volunteer prior to involvement in the experiment (Check Appendix B).

Environmental Issues: There were no environmental issues to be taken into account.

3. Analysis

3.1 Data Collection

3.1.1 Qualitative Data

Table 1: Observed color of saliva assay solutions (exposed to different concentrations of VG (%) in e-liquid) after the addition of potassium iodide solution (color reagent).

VG Concentration/ %	Change of Color in Saliva Assay Sample									
	1	2	3	4	5	6	7	8	9	10
0	Orange-yellow	Orange-yellow	Rose-orange	Rose-orange	Orange-yellow	Deep-orange	Rose-orange	Orange-yellow	Deep-orange	Rose-orange
50	Deep-orange	Deep-orange	Rose-orange	Deep-orange	Deep-orange	Rose-orange	Rose-orange	Orange	Orange	rose-orange
60	Rose-orange	Rose-orange	Red-brown	Deep-orange	Red-brown	Deep-orange	Red-brown	Deep-orange	Red-brown	Rose-orange
70	Rose-brown	Red-brown	Red-brown	Rose-brown	Rose-brown	Brown	Brown	Rose-brown	Red-brown	Red-brown
80	Brown-violet	Violet	Brown-violet	Brownish-violet	Violet	Brownish-violet	Brown-violet	Violet	Violet	Brownish-violet

3.1.2 Quantitative Data

Table 2: Spectrophotometric light absorbance reading (AU) of maltose for 10 samples exposed to each VG concentration in e-liquid (%).

VG Concentration/ %	Light Absorbance Reading of Assay Solution/ AU ($\pm$ 0.01) in Different Saliva Samples									
	1	2	3	4	5	6	7	8	9	10
0	0.82	0.84	0.80	0.79	0.83	0.85	0.79	0.84	0.85	0.79
50	0.70	0.71	0.75	0.72	0.73	0.74	0.75	0.76	0.75	0.74
60	0.62	0.64	0.60	0.65	0.59	0.63	0.61	0.66	0.58	0.62
70	0.44	0.51	0.53	0.39	0.45	0.40	0.40	0.49	0.52	0.50
80	0.24	0.30	0.26	0.29	0.32	0.27	0.26	0.31	0.30	0.27

The uncertainty of the light absorbance reading was calculated by finding the smallest reading division the spectrophotometer could give; which was 0.01. As for the VG concentration (%), no uncertainty could be found considering its value was given on the label of each volunteer's e-liquid.

3.2 Data Processing
3.2.1 Measures of Spread

To understand the relation between VG concentration (%) and the light absorbance reading of maltose found using the spectrophotometer, the means of the 10 saliva assay solutions for each concentration of VG (%) was calculated through the following equation:

$$Mean = \frac{\sum samples\ absorbance}{number\ of\ samples}$$

For example, the set of samples exposed to 0%VG had a mean of: (0.82 + 0.84 + 0.80 + 0.79 + 0.83 + 0.85 + 0.79 + 0.84 + 0.85 + 0.79) / 10 = 0.82.

Furthermore, the standard deviation of each data set was calculated through the following equation:

$$\sigma = \sqrt{\frac{\sum (x_i - \mu)^2}{n - 1}}$$

For example, the same set of samples had a standard deviation of:

97

$$\frac{(0.82-0.82)^2+(0.84-0.82)^2+(0.80-0.82)^2+(0.79-0.82)^2+(0.83-0.82)^2+(0.85-0.82)^2+(0.79-0.82)^2+(0.84-0.82)^2+(0.85-0.82)^2+(0.79-0.82)^2}{10}$$

$$= 0.03 \ (rounded\ to\ 2\ decimal\ points)$$

To determine whether the standard deviation is small or large, the coefficient of variation (CV) of each data set was calculated through the following equation:

$$CV = \frac{\sigma}{\mu} * 100\%$$

For example, the same set of samples had a coefficient of variation of: (0.03)/(0.82)*100 = 3.65% (rounded to 2 decimal points).

Lastly, when each of the above equations were applied to each of the other data sets, the following table was constructed.

Table 3: The mean and spread of the maltose absorbance readings in saliva assay samples exposed to different VG concentrations (%) in e-liquid.

VG Concentration/ %	Mean Maltose Absorbance/ AU	Standard Deviation/ AU	Coefficient of Variation/ %
0	0.82	0.03	3.66
50	0.74	0.01	1.35
60	0.62	0.02	3.23
70	0.46	0.05	10.87
80	0.28	0.03	10.74

The mean reading, standard deviation and coefficient of variation values were all rounded to 2 decimal places, as the absorbance reading values in Table 1 were found in 2 decimal places.

When the VG concentration (%) and mean data was plotted on to an X-Y plane, the following graph was obtained:

Graph 1: Effect of using different concentrations of VG (%) in e-liquid on the light absorbance reading of maltose (AU) in saliva samples.

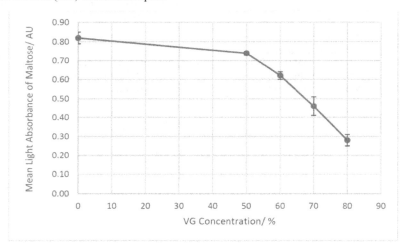

The error bars in the above graph represent the standard deviation found in Table 3.

3.2.2 T-test

Nevertheless, 10 t-tests were performed on the data from Table 1 to establish a statistical difference between the 5 sample sets.

H_0 **(null hypothesis):** There is no statistically significant difference between the mean light absorbance readings of the samples exposed to different VG concentrations (0%, 50%, 60%, 70%, 80%) in e-liquid.

H_1: There is a statistical significant difference between the mean light absorbance readings of the samples exposed to different VG concentrations (0%, 50%, 60%, 70%, 80%) in e-liquid.

Table 4: T-test performed between the 5 saliva sample sets; which each exposed to a different VG concentration (%).

Concentration of VG (%)	T-Calculated	Accept or Reject Alternative Hypothesis (95% of Confidence Level)
0 and 50	8.00	Accepted
0 and 60	17.54	Accepted
0 and 70	19.57	Accepted
0 and 80	40.30	Accepted
50 and 60	16.90	Accepted
50 and 70	17.39	Accepted
50 and 80	46.00	Accepted
60 and 70	9.41	Accepted
60 and 80	29.82	Accepted
70 and 80	9.78	Accepted

Calculation of t-value between every two VG concentrations (%) in e-liquid:

- T-Critical value taken from literature table (Check Appendix C), T-Critical = 2.101

- P-Value = 0.05 (95% confidence level)

- Degree of freedom: $df = (n_1 - 1) + (n_2 - 1) = (10\text{-}1) + (10\text{-}1) = 18$

- T-Calculated: calculate using TI-Nspire CX graphing calculator.

- The T-Calculated in each t-test is > 2.101 (the T-Critical); therefore, we accept the alternative hypothesis at 95% confidence level because the T-Calculated is higher than the T-Critical.

3.2.3 Discussion

In table 1, the increased disappearance of the shades of violet colors to that of shades of orange as the VG concentration (%) decreased indicated higher salivary α-amylase enzyme activity, as more starch was hydrolyzed into maltose. This was quantitatively observed in table 2, where the samples' absorbance readings of maltose (AU) increased as the VG concentration (%) decreased as well; as evident in each set of samples' mean values found in table 3. As shown in graph 1, the mean maltose absorbance readings (AU) underwent a gradual decrease of only 0.08 AU between 0% and 50% VG; however, as the VG concentration (%) increased there was a further decline of 0.12 AU, 0.16 AU, and 0.18 AU respectively in the readings. Further, the standard deviation error bars in the graph were small; suggesting low variation in the maltose absorbance readings (AU) of samples exposed to each VG concentration (%). This was evident in the results of their coefficient of variation shown in table 3; where all their standard deviation values were less than 33% of their sample set's mean. Moreover, it is important to note that none of the error bars overlapped; proposing a potential statistical significant difference between the mean maltose absorbance readings (AU). This potential was made true through the results of the t-tests performed between the 5 saliva assay sample sets; where all the calculated t-values (8.00, 17.54, 19.57, 40.30, 16.90, 17.39, 46.00, 9.41, 29.82, 9.78) were greater than the t-critical value (2.101); thus accepting the alternative hypothesis.

4. Evaluation
4.1 Conclusion

What can be concluded from the results of the experiment is that as the VG concentration (%) in e-liquid increases, 30 μL of its user's salivary α-amylase enzyme activity measured through the spectrophotometric absorbance of maltose (AU) decreases post adding 1.0000 mL of 5% starch solution. In turn, my hypothesis proposing that the increase of VG concentration (%) causes a decrease in the activity of the salivary α-amylase, as less amounts of starch are hydrolyzed into maltose due to the increase in accumulated carbonylated α-amylase enzymes; which is caused by the enzymes' (in the saliva) increased exposure to acrolein gas, is accepted. This explanation of the relationship was supported by literature which stated that VG, when heated to temperatures of 180°C or greater (Stevens & Maier, 2008) – all smoking saliva sample volunteers had their e-cigarettes set at a vaporizing temperature of 250 °C (see 2.6: subject inclusion criteria) –, produced acrolein gas (Ogunwale et al., 2017) which caused the salivary α-amylase enzyme to undergo carbonylation when exposed to it (Nagler et al., 2000). This was first noticed in the qualitative data collected (see table 1), where considering that the amount of maltose left in each saliva assay solution's after the hydrolysis of starch is an indicator of the saliva sample's salivary α-amylase activity (Bagchi & Nair, 2018), the gradual disappearance of the blue-black (or shades of violet in the case of this experiment in particular) color (which iodine turns the solution into in the presence of starch) (Thomas & Atwell, 1999 as cited in Nwokocha & Ogunmola, 2014) as the VG concentration (%) decreased indicated a higher α-amylase activity in the samples. Then, it was noticed in the quantitative data collected (see table 2), where considering that according to the Beer-Lambert law stating that there is a linear relationship between the spectrophotometric absorbance (AU) of the analyte of interest (maltose) and its concentration in the sample (saliva assay solution) (Wypych, 2018), saliva samples exposed to lower VG concentrations (%) had higher maltose absorbance readings (AU); thus indicating higher α-amylase activities due to their readiness in hydrolyzing starch into maltose. This was shown in graph 1 where a negative relationship between the increase of VG concentration (%) in e-liquid and the mean light absorbance of maltose (AU) in the saliva assay solutions was evident; indicating that the amount of maltose in the saliva samples decreased as the concentration of VG (%) it is exposed to increased.

The standard deviation and coefficient of variation of the mean maltose absorbance readings (AU) in each set of samples exposed to a different VG concentration (%) were found to be low, signifying that the variance is low among the maltose absorbance readings of the same set. The statistically significance difference between the 5 sets of samples was initially suggested through graph 1's error bars that did not overlap, and established through the rejection of the null hypotheses in the t-tests performed between them (totaling 10 t-tests).

4.2 Strengths

Since the investigation was conducted using advanced measuring technologies, the percentage uncertainty carried with the results was low, for the volumes of saliva and starch used in the salivary α-amylase assay of each saliva sample had an uncertainty of 0.9% and 0.6%, respectively (totaling 1.4%). This increased the reliability of the collected data. Moreover, the low standard deviation and coefficient of variation values of each set of saliva samples demonstrated a low chance of error (meaning the range of data in each set was significantly close to its mean); which increased the precision of the processed data. Lastly, the number of the controlled variables and the level at which they were controlled helped obtain accurate results.

4.3 Weaknesses

First, only 4 concentrations of VG in e-liquid (50%, 60%, 70%, 80%) were tested, although there are more commercially available concentrations; in turn slightly decreasing the reliability of the conclusion made about the investigation. However, as previously mentioned – it was extremely difficult to find users whom smoked concentrations out of this range; as it was the common one. Second, the experiment design could have included the construction of a maltose standard calibration curve, which would have allowed for the determination of the numerical value of maltose found in each saliva assay solution after the hydrolysis of starch; instead of depending on the spectrophotometric maltose absorbance reading (AU) in accordance to the Beer-Lambert law. However, it is usually constructed using 3,5-Dinitro salicylic acid (DNS) as the color reagent ("Construction of Maltose Standard Curve by DNS Method," 2011) instead of the one used in this experiment (potassium iodide); which was used because it is the most reliable and common detector of the disappearance of starch used in literature. Third, although the total percentage uncertainty of the measuring tools by which the saliva samples and the starch solution were measured was low; it still serves as a systemic error which may have slightly decreased the accuracy of the collected data. Fourth, considering the relatively small sample size tested (10 samples), a random error can be derived from the previous weakness; as the uncertainty of the processed data will increase. And fifth, since the time duration between sample collection and α-amylase was not precisely controlled, there was a slight difference in the environmental conditions each sample went under; which might cause for an inequality in the data collected.

4.4 Extensions

In addressing of each weakness: First, the full range of the commercially available VG concentration (from 10% to 100% VG) could be tested by extensively trying to find more volunteers. However, it might be unrealistic as it would make the investigation significantly more difficult and time-consuming. Therefore, an alternative solution would be to in vitro expose e-cigarette vapor using an e-cigarette smoke-machine to healthy non-smoking saliva samples then test their absorbance. Second, the DNS color reagent (with its corresponding experiment design) could be used instead of potassium iodide; in turn increasing the precision of the processed data on which the conclusion of the investigation is dependent on. Third, measuring tools of lower uncertainties could be used. Fourth, a larger sample size could be tested to decrease the possibility of any uncertainties to interfere with the reliability of the conclusion. And fifth, the single uncontrolled variable (although monitored) could be controlled by marking the time each sample was collected and executing its assay at an exact time period after (preferably no more than a few hours).

Nwokocha, L. M., & Ogunmola, G. B. (2014). Colour of starch-iodine complex as index of retrogradability of starch pastes. *African Journal of Pure and Applied Chemistry*, *8*(5), 89–93. doi: 10.5897/ajpac2014.0571

Ogunwale, M. A., Li, M., Raju, M. V. R., Chen, Y., Nantz, M. H., Conklin, D. J., & Fu, X.-A. (2017). Aldehyde Detection in Electronic Cigarette Aerosols. *ACS Omega*, *2*(3), 1207–1214. doi: 10.1021/acsomega.6b00489

Phillips, B., Titz, B., Kogel, U., Sharma, D., Leroy, P., Xiang, Y., … Vanscheeuwijck, P. (2017). Toxicity of the main electronic cigarette components, propylene glycol, glycerin, and nicotine, in Sprague-Dawley rats in a 90-day OECD inhalation study complemented by molecular endpoints. *Food and Chemical Toxicology*, *109*, 315–332. doi: 10.1016/j.fct.2017.09.001

Rahman, M., Hann, N., Wilson, A., & Worrall-Carter, L. (2014). Electronic cigarettes: patterns of use, health effects, use in smoking cessation and regulatory issues. *Tobacco Induced Diseases*, *12*(1), 21. doi: 10.1186/1617-9625-12-21

Ros, J. (2017). *Protein carbonylation: principles, analysis, and biological implications*. Hoboken, New Jersey: Wiley.

Rudeekulthamrong, P., & Kaulpiboon, J. (2012). Kinetic Inhibition of Human Salivary Alpha-Amylase by a Novel Cellobiose-Containing Tetrasaccharide. *Journal of the Medical Association of Thailand*, S102–S108.

Scannapieco, F. A., Bergey, E. J., Reddy, M. S., & Levine, M. J. (1989). Characterization of salivary alpha-amylase binding to Streptococcus sanguis. Infection and Immunity, 57(9), 2853–2863. doi: 10.1128/iai.57.9.2853-2863.1989

Spectrophotometry. (2019, September 30). Retrieved February 25, 2020, from https://chem.libretexts.org/Bookshelves/Physical_and_Theoretical_Chemistry_Textbook_Maps/Supplemental_Modules_(Physical_and_Theoretical_Chemistry)/Kinetics/Reaction_Rates/Experimental_Determination_of_Kinetcs/Spectrophotometry

Stevens, J. F., & Maier, C. S. (2008). Acrolein: Sources, metabolism, and biomolecular interactions relevant to human health and disease. *Molecular Nutrition & Food Research*, *52*(1), 7–25. doi: 10.1002/mnfr.200700412

The Editors of Encyclopaedia Britannica. (2019, August 16). Glycerol. Retrieved February 25, 2020, from https://www.britannica.com/science/glycerol#ref26585

Tiwari, M. (2011). Science behind human saliva. *Journal of Natural Science, Biology and Medicine*, *2*(1), 53. doi: 10.4103/0976-9668.82322

Vaping Devices (Electronic Cigarettes). (2020, January 8). Retrieved March 9, 2020, from https://www.drugabuse.gov/publications/drugfacts/vaping-devices-electronic-cigarettes

Weiner, D., Levy, Y., Khankin, E. V., & Reznick, A. Z. (2009). Inhibition of salivary amylase activity by cigarette smoke aldehydes. *Journal of Physiology and Pharmacology*, *59*(6), 727–737.

Wypych, G. (2018). Photophysics. *Handbook of Material Weathering*, 1–26. doi: 10.1016/b978-1-927885-31-4.50003-0

Wong, C.-M., Bansal, G., Marcocci, L., & Suzuki, Y. J. (2012). Proposed role of primary protein carbonylation in cell signaling. Redox Report, 17(2), 90–94. doi: 10.1179/1351000212y.0000000007

References

A law gone up in smoke. (2017, July 2). Retrieved February 25, 2020, from https://jordantimes.com/opinion/editorial/law-gone-smoke

AccuWeather. (n.d.). Retrieved February 25, 2020, from https://www.accuweather.com/en/jo/amman/221790/october-weather/221790?year=2019

Acrolein. (n.d.). Retrieved February 25, 2020, from https://pubchem.ncbi.nlm.nih.gov/compound/Acrolein.

Bagchi, D., & Nair, S. (2018). *Nutritional and therapeutic interventions for diabetes and metabolic syndrome* (2nd ed.). London: Academic Press.

Barranco, T., Tvarijonaviciute, A., Escribano, D., Tecles, F., Cerón, J. J., Cugat, R., ... Rubio, C. P. (2018). Changes of salivary biomarkers under different storage conditions: effects of temperature and length of storage. *Biochemia Medica*, 29(1), 94–111. doi: 10.11613/bm.2019.010706

Bekki, K., Uchiyama, S., Ohta, K., Inaba, Y., Nakagome, H., & Kunugita, N. (2014). Carbonyl Compounds Generated from Electronic Cigarettes. *International Journal of Environmental Research and Public Health*, 11(11), 11192–11200. doi: 10.3390/ijerph111111192

Brueck, H. (2019, November 15). The wild history of vaping, from a 1927 'electric vaporizer' to today's mysterious lung injury crisis. Retrieved February 25, 2020, from http://www.insider.com/history-of-vaping-who-invented-e-cigs-2019-10#i-believed-that-if-i-could-use-vapor-to-simulate-cigarette-smoke-this-could-help-me-lik-said-9.

Caprette, D. R. (1997, May 8). Selected Critical Values of the t-Distribution. Retrieved March 9, 2020, from https://www.ruf.rice.edu/~bioslabs/tools/stats/ttable.html

Caraway, W. T. (1959). A Stable Starch Substrate for the Determination of Amylase in Serum and Other Body Fluids. *American Journal of Clinical Pathology*, 32(1_ts), 97–99. doi: 10.1093/ajcp/32.1_ts.97

Cassidy, S. (2020, January 27). How Electronic Cigarettes Work. Retrieved February 25, 2020, from https://science.howstuffworks.com/innovation/everyday-innovations/electronic-cigarette1.htm

Cheesbrough, M. (1999). *District laboratory practice in tropical countries*. Cambridge, United Kingdom: Cambridge University Press.

Construction of Maltose Standard Curve by DNS Method. (2011). Retrieved February 27, 2020, from http://vlab.amrita.edu/?sub=3&brch=64&sim=163&cnt=1

Devito, E. E., & Krishnan-Sarin, S. (2018). E-cigarettes: Impact of E-Liquid Components and Device Characteristics on Nicotine Exposure. *Current Neuropharmacology*, 16(4), 438–459. doi: 10.2174/1570159x15666171016164430

Jones, L. (2019, September 15). Vaping: How popular are e-cigarettes? Retrieved February 25, 2020, from https://www.bbc.com/news/business-44295336

Kosmider, L., Sobczak, A., Fik, M., Knysak, J., Zaciera, M., Kurek, J., & Goniewicz, M. L. (2014). Carbonyl Compounds in Electronic Cigarette Vapors: Effects of Nicotine Solvent and Battery Output Voltage. *Nicotine & Tobacco Research*, 16(10), 1319–1326. doi: 10.1093/ntr/ntu078

Lai, O.-M., Tan, C.-P., & Akoh, C. C. (Eds.). (2012). *Palm oil: Production, Processing, Characterization, and Uses*. Urbana, Illinois: AOCS Press.

Mandel, A. L., & Breslin, P. A. S. (2012). High Endogenous Salivary Amylase Activity Is Associated with Improved Glycemic Homeostasis following Starch Ingestion in Adults. *The Journal of Nutrition*, 142(5), 853–858. doi: 10.3945/jn.111.156984

Moore, R. J., Watts, J. T. F., Hood, J. A. A., & Burritt, D. J. (1999). Intra-oral temperature variation over 24 hours. *The European Journal of Orthodontics*, 21(3), 249–261. doi: 10.1093/ejo/21.3.249

Nagler, R., Lischinsky, S., Diamond, E., Drigues, N., Klein, I., & Reznick, A. Z. (2000). Effect of Cigarette Smoke on Salivary Proteins and Enzyme Activities. *Archives of Biochemistry and Biophysics*, 379(2), 229–236. doi: 10.1006/abbi.2000.1877

Nakajima, K. (2016). Low serum amylase and obesity, diabetes and metabolic syndrome: A novel interpretation. *World Journal of Diabetes*, 7(6), 112. doi: 10.4239/wjd.v7.i6.112

Nilapwar, S. M., Nardelli, M., Westerhoff, H. V., & Verma, M. (2011). Absorption Spectroscopy. *Methods in Enzymology Methods in Systems Biology*, 500, 59–75. doi: 10.1016/b978-0-12-385118-5.00004-9

6. TO WHAT EXTENT DOES THE SAP OF THE HERACLEUM MANTEGAZZIANUM CAUSE BURN MARKS (CM2) ON THE LEAVES OF FIVE DIFFERENT PLANT TYPES

Author: Yara Alzubi
Moderated Mark: 21/24

Introduction:

Research question:

"To what extent does the sap of the <u>Heracleum mantegazzianum</u> cause burn marks (cm²) on the leaves of five different plant types in terms of the percentage they cover in the first 24 hours after application?"

Background information:

The *Heracleum mantegazzianum (H. mantegazzianum)* plants, also known as Giant hogweed, have been on an official "dangerous plants list" for over 45 years (CBS news). When prepared correctly, it can be used as treatment for a sore throat, swelling, cuts and muscle sprains (Herbpathy). However, besides it being a very invasive plant species (RHS), it is known to cause severe blisters and pain when its sap comes into contact with human skin, because it contains organic chemicals called furocoumarins (Perrone). These chemicals absorb the long-wave ultraviolet radiation from sunlight, which causes a damaging reaction resulting in cell damage (Lawley). The plants seem to originate from Russia (CBS News), but are now regularly seen throughout Europe and the US (Perrone), as they spread by seed (RHS). The plants, which can be taller than three meters, are especially a threat during the summer (RHS). Due to this, crews from the state Department of Environmental Conservation have been cutting out *H. mantegazzianum* plants in public areas for years now (Nearing).

Relevance:

The sap its effects on human skin are well known, however not much research has been done on the effects of the sap on other plants. As the sap inhibits DNA synthesis (Lawley), damage to the plant leaves is inevitable once the two come in contact. This could, for example, happen when the stem of a *H. mantegazzianum* snaps due to a burst of wind, but also when the plant is not removed correctly. The plants that are damaged could be part of a food chain or have a different purpose in the environment. When a *H. mantegazzianum* snaps or is not removed properly, chances of exposure to the sap increase for these plants. This experiment will try to show the effects when this happens, proving that precautions for the *H. mantegazzianum* plants must not only be taken for humans sake, but also for that of the plants growing in the same environment.

Personal Engagement:

I live in an area with huge fields of *H. mantegazzianum* plants (picture 1). One of my neighbours once fell into them, causing sap of the plant to come in contact with her arm. The blisters were huge. This made me wonder how dangerous they are for their surrounding plants and whether they also caused burn marks on them, especially now that the government is working hard to remove the *H. mantegazzianum* plants. I know the experiment is dangerous, but I asked my grandfather, who is a certified gardener and has removed many of the plants throughout his life, to supervise me while I conduct my method.

Picture 1: Field of <u>H. mantegazzianum</u> plants

Hypothesis:

Aim:

The aim of this experiment is to show that the sap of the *H. mantegazzianum* negatively affects some plant types that grow in the same area by making a ranking, from plants that are most affected to least affected by the sap.

Hypothesis:

The expectation is that the sap of the *H. mantegazzianum* in some way will affect all plants. This is because all leaves contain stomata, which are pores that absorb carbon dioxide in order to photosynthesize, and guard cells, which are in charge of a process called transpiration (Allot). Transpiration is the rate that gas is absorbed while water vapour is released and exerted from the leaves. When the stomata and guard cells are damaged, gas exchange can no longer occur efficiently and too much water can be lost, which can result in damage to the leaves (Allot). The leaves of the plant types that are tested are not used to the acidity of the sap, causing it to likely burn through the stomata and the guard cells.

The burn marks on the leaves will especially grow during the daylight hours, because the furocoumarins in the sap require UV light in order to react with the nitrogenous DNA bases of the leaves (Brunning, Hogweed).

It is likely that the sap causes the greatest burn mark cover percentage on the *P. geraniums, Solidago*s and the *Hydrangea*s, as they have the thinnest leaves. The *R. ferrugineums* have much thicker leaves, making it less likely that bigger burn marks start to appear. It is expected that the *U. dioicas* will have the smallest burn marks, because the plant already stores acidic sap by itself (Brunning, *Urtica dioicas)*, making it seem likelier that it is more immune to the highly acidic sap of the *H. mantegazzianum*.

Method:

Variables:

Independent variables	Which ones?	Why an independent variable?	Uncertainty
The five plant types	1. *Pelargonium geranium* (*P. geranium*) 2. *Solidagos* 3. *Rhododendron ferrugineum* (*R. ferrugineums*) 4. *Hydrangeas* 5. *Urtica dioicas* (*U. dioicas*)	These types grow in the environment of the *H. mantegazzianum* and vary in their appearance, leaf contents, leaf thickness, and growth behaviour, making them good representations of all the plants that could be affected. As they are all different from each other to an extent, sufficient data can be found, proving the effect of the sap on different plant conditions.	No uncertainty, as no plants are hybrids.

Dependent variables	Why a dependent variable?	Uncertainty
Size of the burn marks (cm^2) and the percentage of the leave burned	They depend on the type of plant leaf, as, for example, some plants have thicker leaves than others. The thicker and stronger leaves are likely less affected by the sap. There will be five trials for each plant type, making it less likely that inaccuracies such as one leaf getting more sun than the others will affect the data. This increases data accuracy and reliability.	There is an uncertainty of $0.05\ cm^2$ and 0.05%

Controlled variables	Why controlled?	Method of control	Uncertainty
The size of the leaves used when calculating the burn mark percentage	Each leaf falls under at least one of the size categories, which are Large (L), medium (M) and small (S). The corresponding amount of juice is applied to the leaf, so that the size to juice ratio stays controlled and the experiment is likely to be more reliable.	Before the experiment, the leaves are measured using a ruler. Each leaf should fit in one of the three size categories described. Their actual size is used when calculating the burn mark percentage.	There is an uncertainty of 0.05 cm^2.
Amount of sap applied	If one plant gets more sap than the other, the ranking of the plants that are most to least affected will be inaccurate, because the burn marks will likely be bigger when more sap is applied.	2 ml of sap is applied on each large leaf with a surface area of 58-66 cm^2, 1 ml to medium sized leaves of 28-35 cm^2, and 0.5 ml to small sized leaves of around 11-15 cm^2. This is done using kitchen measuring spoons.	There is an uncertainty of 0.05 ml.
The weather conditions the plants are in	Weather conditions such as rain, wind and snow can cause harm to the plants. Sunlight affects the rate of photosynthesis of the plants.	The experiment is done during a dry, slightly cloudy summer day, so no damage is done by extreme weather factors. All the plants are tested at the same time, so their weather conditions are the same.	No uncertainty.
Length of the experiment and moment data is collected	The acidic sap has most effect in the first couple of hours and one day will probably give a clear representation of its effect on the plants. Data is collected at the exact same time intervals, so that an accurate ranking can be made of the plants from most affected to least.	The experiment takes 24 hours. A timer goes off after 15 minutes, 1 hour, 12 hours, 18 hours, and 24 hours, to help remember. At those moments, pictures are taken at several angles of the leaves, to later be analyzed for the size of the burn marks.	There is an uncertainty of about two minutes. Pictures are taken, instead of measuring on the spot, as the time uncertainty would then be too big.

Safety and risk assessment:

This experiment can be dangerous if no precautions are taken. Wear thick, gardening gloves and goggles at all times. Make sure you are wearing a long sleeved shirt, long pants, socks and closed shoes, so that no skin is exposed. Ask an expert to supervise while conducting the experiment. Throw out all the used equipment once the experiment is over, taped in plastic or paper, so that no one can accidently get in touch with it directly. Handle the sap with care, making sure it does not get in contact with any plants that are not part of the experiment. Never take risks and stay alert at all times. If skin is accidently exposed to the sap anyway, thoroughly rinse the area with water and cover it with a special ointment for burns.

<u>List of materials:</u>
- Gloves and goggles
- Ruler of 30 centimetres   (uncertainty of 0,05 cm)
- One *H. mantegazzianum* between 1,5 to 3 meters tall
- Five *P. geranium, Solidago, R. ferrugineums, Hydrangea,* and *U. dioica* leaves of each
- Twenty-five ribbons
- Twenty-five 5 cm x 5 cm plastic sheets
- 0.5 ml, 1 ml, and 2 ml measuring spoons   (uncertainty of 0,05 ml in all cases)
- Pestle and mortar
- Camera
- Hedge cutter
- Timer   (uncertainty of 0,005 seconds)
- See-through paper, pencil and highlighter
- Scissors
- Plastic bag

<u>Methodology:</u>
1. Measure all the leaves using the ruler. Decide which leaf falls under which size category. Mark them with a ribbon and take pictures. Write down their actual sizes for later use.
2. Put on your gloves and goggles. Keep them on at all times. Take the hedge cutter and cut a 10 centimetre long stem of the *H. mantegazzianum,* using a ruler, found in the nearest woods or field. Put it in the plastic bag and take it to a safer place before taking the sap out.
3. Cut the stem in smaller pieces and put it in the mortar. Carefully beat the stem until a mush/sap starts to form, using the pestle. Get the twenty-five plastic 5x5 sheets and the measuring spoons. Apply the appropriate amount of the sap on each plastic sheet and label them accordingly.
4. On each leaf, rub one plastic sheet with the appropriate amount of sap on it, over the entire surface of the leave. This should be done to five leaves of each plant type. In this case, this was done in June at 9:00 PM. This time makes it the easiest to collect data at all timeframes.
5. Take a picture of the leaves after 15 minutes, 1 hour, 12 hours, 18 hours, and 24 hours, from all angles, a ruler distance above the plants. Do not zoom in. Set a timer for these times to remind you.
6. After finishing, throw away all the equipment used. Wash your hands thoroughly with water while still wearing the gloves. Carefully take them off and throw them away.
7. For each leaf, choose the clearest picture from any angle at each timeframe and print the pictures on high resolution.
8. Get the see-through paper, the photos, pencil and highlighter, and stand in front of a window. Put the picture on the window with the see-through paper over it, and draw the leaves and their burn marks. Take your time for this. Name each drawing correctly.
9. Once all drawings are finished, calculate the area of the burn marks on each leaf. Sometimes it is easier to cut the burn marks out, using a scissor, and put them together as a square. Write the values down in cm^2. Calculate the percentage of the leaf that was burned after each time frame and write these values down too.

Raw data: qualitative

Figure 1: Table with all observations that were made during the experiment for each plant type at each timeframe.

	U. dioicas	R. ferrugineums	P. geraniums	Solidagos	Hydrangeas
15 min. (± 2 min)	Started to show burn marks.	Yellowish border around leaves, which curled up.	Burn marks were clearly visible.	No sign of any burn marks.	Yellowish border around leaves, which curled up.
1 h. (± 2 min)	Seemed to have slightly bigger burn marks.	Leaves seemed similar to the 15-minute time mark.	Leaves seemed similar to the 15-minute time mark.	Tiny ruptures in the centre of the leaves.	Seemed to have slightly bigger burn marks.
12 h. (± 2 min)	Holes in some of the leaves	Slightly bigger burn marks.	Bigger burn marks.	Bigger ruptures	Holes in some of the leaves
18 h. (± 2 min)	Brown marks and holes in its leaves.	Burn marks on the bottom of some leaves.	Big brown marks and spots around the corners and in the centre of the leaves.	No brown marks. Ruptures were slightly larger.	Big brown marks and holes around the corners and in the centre of the leaves.
24 h. (± 2 min)	Darker burn marks, so more were visible. Reaction seemed to be over (Picture 5).	No big changes visible. Reaction seemed to be over.	Darker burn marks, so more were visible. Reaction seemed to be over.	No big changes visible. Reaction seemed to be over	Darker burn marks, so more were visible. Reaction seemed to be over.

Picture 2: <u>Hydrangea</u> leaves before experiment

Picture 3: <u>Hydrangea</u> leaf #2 after 24 hours

Picture 4: <u>U. dioica</u> leaves before experiment

Picture 5: <u>U. dioica</u> leaf #4 after 24 hours

Raw data: quantitative

Figure 2: Table with all raw data for each leaf of each plant type after each timeframe.

	Size categ- ory	Actual size in cm^2 (± 0.05 cm^2)	Burned area in cm^2 (± 0.05 cm^2) 15 min. (± 2 min)	Burned area in cm^2 (± 0.05 cm^2) 1 h. (± 2 min)	Burned area in cm^2 (± 0.05 cm^2) 12 h. (± 2 min)	Burned area in cm^2 (± 0.05 cm^2) 18 h. (± 2 min)	Burned area in cm^2 (± 0.05 cm^2) 24 h. (± 2 min)
U. dioica #1	S	11.1	0.4	1.4	1.7	3.3	4.3
U. dioica #2	M	32.3	1.9	3.2	5.5	7.1	7.5
U. dioica #3	M	30.5	1.3	2.9	7.4	8.0	8.5
U. dioica #4	S	13.9	0.2	2.0	2.9	3.7	4.7
U. dioica #5	S	14.3	0.8	2.5	4.0	4.9	5.1
R. ferrugineum #1	M	34.1	0.0	0.0	0.4	0.8	0.9
R. ferrugineum #2	M	32.8	0.0	0.0	1.2	2.5	2.6
R. ferrugineum #3	M	28.6	0.0	0.0	0.9	2.3	2.3
R. ferrugineum #4	M	29.0	0.0	0.0	0.8	1.3	1.6
R. ferrugineum #5	M	27.6	0.0	0.0	1.5	2.4	2.9
P. geranium #1	L	62.3	14.9	15.3	30.3	34.7	36.1
P. geranium #2	L	58.8	14.5	15.9	23.3	25.5	26.8
P. geranium #3	M	31.2	4.6	5.0	10.4	13.2	14.5
P. geranium #4	M	28.4	3.7	3.9	10.9	12.9	13.7
P. geranium #5	M	29.0	3.9	4.3	12.1	14.1	15.6
Solidago #1	S	13.1	0.0	0.0	0.4	0.6	0.8
Solidago #2	S	12.7	0.0	0.2	0.7	0.7	0.9
Solidago #3	S	13.6	0.0	0.3	0.7	0.9	1.0
Solidago #4	S	13.3	0.2	0.5	0.9	1.1	1.2
Solidago #5	S	11.4	0.0	0.0	0.2	0.5	0.6
Hydrangea #1	L	64.7	0.4	3.3	28.9	34.6	37.5
Hydrangea #2	L	61.3	0.6	5.6	33.1	38.1	46.4
Hydrangea #3	L	66.0	0.0	3.7	26.4	29.5	29.8
Hydrangea #4	L	62.0	0.0	2.4	32.0	35.2	36.7
Hydrangea #5	L	59.9	0.0	4.1	31.4	33.9	34.3

Processed data

Using the raw data in figure 2, the percentage of the leaves that is covered with burn marks was calculated in the following way: as an example, leaf 1 from the *P. geraniums* is taken as an example. The leaf was big, with an actual size of 62.3 cm^2. After 15 minutes, an area of 14.9 cm^2 on the leaf was burned. To calculate the percentage, the area that was burned is divided by the actual size of the leaf, so $\frac{14.9}{62.3} \times 100 = 23.9\%$. Using this method, the percentage of the leaves that is covered with burn marks was calculated for each leaf, after each timeframe. This resulted in the data found below (figure 3-7).

These values were calculated, because it would otherwise be much more complicated to compare the raw data with each other. This would make it impossible to rank the plant types based on which ones are most affected by the sap of the *H. mantegazzianum*, and it would be harder to comprehend to what extent the leaves were burned in general. However this way, the data is comprehendible, understandable, and easy to draw conclusions from.

Figure 3: Data of the *U. dioica* leaves. There is an uncertainty of 0.05% in all cases.

	Percentage burned after 15 min. (± 2 min)	Percentage burned after 1 h. (± 2 min)	Percentage burned after 12 h. (± 2 min)	Percentage burned after 18 h. (± 2 min)	Percentage burned after 24 h. (± 2 min)
#1	3.6%	9.9%	15.3%	29.7%	38.7%
#2	5.9%	9.9%	17.0%	22.0%	23.2%
#3	4.3%	9.5%	24.3%	26.2%	27.9%
#4	1.4%	14.4%	20.9%	26.6%	33.8%
#5	5.6%	17.5%	28.0%	34.3%	35.7%

All *U. dioica* leaves already had burn marks after 15 minutes, which grew bigger over time. After 24 hours, leaf 1 was most covered, by 38.7%, with burn marks. All values have an uncertainty of 0.05%.

Figure 4: Data of the *R. ferrugineum* leaves. There is an uncertainty of 0.05% in all cases.

	Percentage burned after 15 min. (± 2 min)	Percentage burned after 1 h. (± 2 min)	Percentage burned after 12 h. (± 2 min)	Percentage burned after 18 h. (± 2 min)	Percentage burned after 24 h. (± 2 min)
#1	0.0%	0.0%	1.2%	2.3%	2.6%
#2	0.0%	0.0%	3.7%	7.6%	7.9%
#3	0.0%	0.0%	3.1%	8.0%	8.0%
#4	0.0%	0.0%	2.8%	4.5%	5.5%
#5	0.0%	0.0%	5.4%	8.7%	10.5%

All *R. ferrugineum* leaves did not show any burn mark after 15 minutes or 1 hour. After this, the leaves developed burn marks, but these did not get very big. After 24 hours, leaf 5 had the biggest burn mark of 10.5%. All values have an uncertainty of 0.05%.

Figure 5: Data of the leaves of the *P. geraniums*. There is an uncertainty of 0.05% in all cases.

	Percentage burned after 15 min. (± 2 min)	Percentage burned after 1 h. (± 2 min)	Percentage burned after 12 h. (± 2 min)	Percentage burned after 18 h. (± 2 min)	Percentage burned after 24 h. (± 2 min)
#1	23.9%	24.6%	48.6%	55.7%	57.9%
#2	24.7%	27.0%	39.6%	43.4%	45.6%
#3	14.7%	16.0%	33.3%	42.3%	46.5%
#4	13.0%	13.7%	38.4%	45.4%	48.2%
#5	13.4%	14.8%	41.7%	48.6%	53.8%

The leaves of the *P. geraniums* already showed burn marks after 15 minutes, which kept growing over time. Leaf 1 showed the biggest burn mark after 24 hours, as it was covered by 57.9%. All values have an uncertainty of 0.05%.

Figure 6: Data of the leaves of the *Solidagos*. There is an uncertainty of 0.05% in all cases.

	Percentage burned after 15 min. (± 2 min)	Percentage burned after 1 h. (± 2 min)	Percentage burned after 12 h. (± 2 min)	Percentage burned after 18 h. (± 2 min)	Percentage burned after 24 h. (± 2 min)
#1	0.0%	0.0%	3.1%	4.6%	6.1%
#2	0.0%	1.6%	5.5%	5.5%	7.1%
#3	0.0%	2.2%	5.1%	6.6%	7.4%
#4	1.5%	3.8%	6.8%	8.3%	9.0%
#5	0.0%	0.0%	1.8%	4.4%	5.3%

Some *Solidago* leaves did not show any burn marks yet after 15 minutes, or even after 1 hour. The burn marks stayed small but did grow slightly over time. Leaf 4 was covered with the biggest burn mark after 24 hours, of 9.0%. All values have an uncertainty of 0.05%.

Figure 7: Data of the *Hydrangea* leaves. There is an uncertainty of 0.05% in all cases.

	Percentage burned after 15 min. (± 2 min)	Percentage burned after 1 h. (± 2 min)	Percentage burned after 12 h. (± 2 min)	Percentage burned after 18 h. (± 2 min)	Percentage burned after 24 h. (± 2 min)
#1	0.6%	5.1%	44.7%	53.5%	58.0%
#2	0.9%	9.1%	54.0%	62.2%	75.7%
#3	0.0%	5.6%	40.0%	44.7%	45.1%
#4	0.0%	3.9%	51.6%	56.8%	59.2%
#5	0.0%	6.8%	52.4%	56.6%	57.3%

Some *Hydrangea* leaves showed little burn mark after 15 minutes, but these grew significantly as time progressed and the sun came up. Leaf 2 had the biggest burn mark of all the leaves that were tested in this experiment after 24 hours, and was covered by 75.7%. All values have an uncertainty of 0.05%.

Once these percentages were found, the averages of the areas burned after a certain time frame per plant type were calculated (figure 8). The averages were calculated, because the amount of values that need to be compared is less, while they are still clear representatives of the effect of the sap on the different types of plants. It is also easier to see patterns within the data values, making it possible to rank the plant types from affected the most to the least. The graph of these averages (figure 9) is especially useful for this.

Figure 8: The average percentages that the leaves are burned. There is an uncertainty of 0.05% in all cases.

	Percentage burned after 15 min. (± 2 min)	Percentage burned after 1 h. (± 2 min)	Percentage burned after 12 h. (± 2 min)	Percentage burned after 18 h. (± 2 min)	Percentage burned after 24. (± 2 min)
U. dioicas	4.2%	12.2%	21.1%	27.8%	31.9%
R. ferrugineums	0.0%	0.0%	3.2%	6.2%	6.9%
P. geraniums	17.9%	19.2%	40.3%	47.1%	50.4%
Solidagos	0.3%	1.5%	4.5%	5.2%	6.9%
Hydrangeas	0.3%	6.1%	48.5%	54.8%	59.1%

113

As the table shows, the averages increase over time for each plant type. The highest averages were either from the *P. geraniums* or the *Hydrangeas*. There is an uncertainty of 0.05% in all cases. A graph of these values was made (figure 9).

Figure 9: Graph of the average percentages that the leaves are burned. There is an uncertainty of 0.05% in all cases.

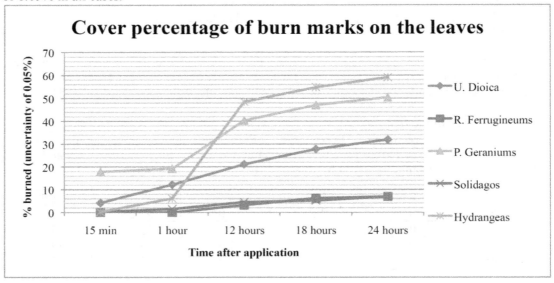

In the graph it is visible that the *R. ferrugineums*, *Solidagos*, and the *U. dioicas*, have quite a constant gradient as time progresses. It is clear that the *R. ferrugineum* and *Solidagos* leaves kept burning on the same rate, coincidentally resulting in the same average percentage burned after 24 hours. After 15 minutes and 1 hour, the *P. geranium* has the biggest burned area on its leaves, however after 12, 18, and 24 hours, the burn marks are the biggest on the *hydrangea* leaves.

Standard deviations
In order to see whether the collected data values can be considered reliable, the standard deviations were calculated using the following formula (Math is fun).

$$\sigma = \sqrt{\frac{1}{N} \sum_{i=1}^{N}(x_i - \mu)^2}$$

It is best if the standard deviations are as small as possible, as this signals that the data values lay close together, are accurate, and to say whether the cover percentages are significantly different amongst the plants. In this case, the standard deviations can also be used to make the most realistic ranking of the five plant types, from being affected the most to the least by the sap of the *H. mantegazzianum*.

114

Figure 10: The standard deviations of the five plant types after each time period.

	15 min	1 hour	12 hours	18 hours	24 hours
P. geraniums	5.2	5.5	5.0	4.8	4.7
Solidagos	0.6	1.1	1.8	1.6	1.3
R. ferrugineums	0.0	0.0	1.4	2.4	2.7
Hydrangeas	0.4	1.8	5.3	5.8	9.8
U. dioicas	1.6	3.2	4.7	4.1	5.6

The standard deviations were calculated using the formula in the following manner. As an example, the calculations of the SD of the *P. geraniums* in the after 15 minutes are shown: First the average of all the values was calculated (figure 8). In this case, this was 17.9%. This average is then subtracted from each data value in the category, so 23.9% − 17.9% = 6%, 24.7% - 17.9% = 6.8%, 14.7% - 17.9% = -3.2%, 13.0 – 17.9% = -4.9%, and 13.4% - 17.9% = -4.5%. These differences are then squared, which results in the numbers, 36, 46.24, 10.24, 24.01, and 20.25. The average of these numbers are calculated by adding them together, which is 136.74, and then divided by 5, the number of data values, so $\frac{136.74}{5} =$ 27.348. The last step is to take the square root of this, resulting in a SD of 5.2.

Unfortunately, the standard deviations are quite high in some cases. The standard deviations can be interpreted in ways that, for example the data values of the leaves of the *P. geraniums* after 15 minutes, are ± 5.2% off from each other. This can make some difference in the ranking that is visible in figure 9. The leaves of the *Hydrangeas* after 24 hours have the highest standard deviation of 9.8.

Possible causes of these varying results can be that some leaves were more in the shadow than the other leaves, which is therefore an uncontrolled variable. The UV light of the sun has great impact on the severity of the burn marks from the sap of the *H. mantegazzianum* (Brunning, Hogweed), so even when a leave stands 10 minutes longer in the shadow than the others, differences can already be seen. Another possibility is that sometimes it was difficult to see whether a spot on a leaf was a burn mark or not. Especially in the first hours after application, some spots could have been missed as they were on the lighter side. Once they got darker they became more visible, however this still resulted in some inaccuracies in the data values.

Data analysis

Looking at figures 3 till 9, it is visible that after 24 hours, the *Hydrangeas* seem to be most affected by the sap of the *H. mantegazzianum*. The *P. geraniums* are affected the second most, followed by the *U. dioicas*. The *P. geraniums* and the *Hydrangeas* had much bigger burn marks after 12 hours than they had after 1 hour, as there was a 21.1% and 42.4% increase on average respectively (figure 8). This is when the plants got more sunlight, so their leaves got more severely damaged. The reason for this is because 95% of the sunlight that comes through the atmosphere has a wavelength of 320-380 nanometers, which activates the furocoumarins in the sap to react with the DNA bases and DNA strands in the stomata and guard cells in the leaves, resulting in cell death and burn marks (Brunning, Hogweed). For the other plant types the reaction was already starting to get nearer to completion, which therefore caused the sunlight to have a much smaller effect.

The cover percentage of burn marks on the *U. dioica* leaves were, on average, always in the middle of the values of the *P. geraniums* and the *Solidagos* (figure 9). It is visible that the leaves of the *U. dioica* behaved differently from the other tested plant types, which shows that the sap of the *H. mantegazzianum* has different effects on different plant species.

The *R. ferrugineums* and *Solidago*s are barely affected and coincidentally both end with an area of burn marks of 6.9% on their leaves, one day after application.

Considering the standard deviations and the actual data values from figure 3-7, instead of just the averages from figure 8, the following ranking can be made, from most to least affected.

First comes the *Hydrangea*, because even though its standard deviations get extremely high as time progressed, almost all leaves are covered with burn mark by more than 50%, one even by 75.7% (picture 3), after 24 hours (figure 7). This last mentioned leaf caused the standard deviation to be so high, as it was an exception to the other leaves, however this does not invalidate the fact that the other leaves were still burned much more compared to leaves of the other plant types. The plant therefore should definitely be ranked first.

Second comes the *Pelargonium*. Its standard deviation was a bit lower towards the end and most leaves were covered in burn marks between 45% and 55% (figure 5), which is lower than the leaves of the *Hydrangea*.

On the third place comes the *U. dioica*. Its data values were right in the middle compared to the other plant types, which is also visible in figure 9, so its standard deviations cannot make a single difference.

On the fourth place comes the *R. ferrugineum*, even though its average was exactly the same with that of the *Solidago*s after 24 hours (figure 8). This is, because almost all leaves are covered with a higher percentage of burn marks by the end of the experiment than the *Solidago* leaves (figure 4 and 6). There was one clear outlier, which was leaf 1 and was only covered by 2.6% (figure 4), possibly due to less sunlight than the other leaves. This lowered the average of the leaves in figure 8 and increased its standard deviation, but overall the leaves were still more affected by the sap than the leaves of the *Solidago*s.

The *Solidago*s come on the fifth and final place. Its standard deviation in the lowest, due to the fact that all leaves were very little affected by the sap. After 24 hours, all leaves were covered with burn marks by less than 9%.

Conclusion

This report therefore shows that the sap of the *H. mantegazzianum* certainly does have an effect on its surrounding plants, just like on human skin, meaning the hypothesis can be accepted to an extent. The plant is clearly extremely dangerous for its surroundings, outweighing its benefits as a medicinal herb. Similar herb treatments for sore throats and swelling exist, without those herbs being so dangerous to their environments. The hypothesis can also be accepted because it correctly predicted that sunlight would have a catalyzing effect on the rate of the reaction. It should be rejected however about the place in the ranking of the *U. dioica*s and *Solidago*s, showing that the thickness or acidity of the leaves are not the only factors that influence the effect of the sap on the plant types. The following ranking can be made of the five tested plant types, from most to the least affected:

1. *Hydrangea*s
2. *P. geranium*s
3. *U. dioica*s
4. *R. ferrugineum*s
5. *Solidago*s

It was thought that the *Hydrangea*s, *P. geranium*s, and *Solidago*s would be most affected as they had the thinnest leaves. This was the case for the first two types, however the *Solidago*s ended up being the least affected by the sap. It turns out that *Solidago*s are immune to ethylene (Reid), which is actually a compound known to prevent cell division or DNA synthesis (Burg). This resembles the consequences of the furocoumarins in the sap, making it

116

possible that the *Solidagos* are therefore also more immune to the effect of the sap compared to other plant types, which can explain why they were affected so little. According to the hypothesis, the *U. dioicas* would be the least affected, because the leaves have acidic hairs themselves, making it more likely that they are used to getting into contact with acidic juices. However, the acid in the hairs of the *U. dioicas*, called trichomes, is quite different, as this contains formic acid, histamine, known to cause rashes and redness, acetylcholine, and many other compounds which are not found in the sap of the *H. mantegazzianum* (Brunning, *Urtica dioicas*). The leaves do seem to be less affected by it compared to the *hydrangeas* and *P. geraniums*, likely because it can withhold a weak acid, but not one as strong as the sap that was applied in this experiment. Lastly, the *R. ferrugineums* were also barely affected, probably because they have such thick leaves, meaning that more sap would be needed before significant burn marks would appear.

To conclude, this experiment shows that the sap of the *H. mantegazzianum* has negative effects on plants that grow in its surroundings in Europe and the US (Perrone), by causing big burn marks, ruptures and cell death inside the leaves. This should be considered while the weeds are being removed, so that less sap can get on the plants. Also, during heavy storms, the *H. mantegazzianums* could fall over, causing spillage of it on the plants around it. Innocent plants, which are part of the food chain of snails for example, get unnecessarily damaged. Every plant loss is one too many, especially during these times where plant diversity is decreasing, and parts of nature are removed or altered, just so that it fits the plans of our rapidly growing cities and industries. As this report shows, there are valid reasons to remove the *H. mantegazzianums*, but it also shows that the other plant types are vulnerable and should be protected from these and similar highly toxic weeds.

Evaluation
The aim of the experiment was met and the desired outcomes were found. The time frame was exactly right, there were enough different conditions to get varying data, and the most essential variables, such as the size and amount of sap per leaf, seemed to have been controlled correctly. There is, however, still some room for improvements. This sample was on the smaller side. Only five plant types and five leaves per type were tested. Therefore, the experiment should be repeated with a bigger sample, at least 25 leaves per plant, in order to really get a clear picture of the negative effects of the sap. This experiment is still reliable, as clear effects were seen, and the chance is statistically very small that this is a coincidence.

The research can also be expanded by, for example, testing the effect of the sap on wood, grass, other weeds, flowers, hay, and etcetera. 24 hours is a good time frame to stick with, as the reaction seemed to almost reach completion.

The method could be more accurate, because it is possible that some human errors were made while drawing and copying the burn marks on the leaves with the see-through paper. The amount of sunlight and shadow that the leaves are in is also essential to the results, so more attention to this is required when the method is repeated. By doing the test in the lab with a UV lamp, these factors can be more controlled for example. Since this was not done like this, the standard deviations are sometimes on the higher side, but luckily it was still possible to make a clear and valid conclusion.

To make the data collection more accurate, ultraviolet photography could be used, as it makes much more detailed and high-resolution pictures. This would make it much easier to see all the burn marks and their actual sizes. A time lapse of the reaction could have also been made, in order to really see at which stage the reaction proceeded the fastest. Unfortunately these methods were not possible for this report due to a lack of the right equipment.

Still, as previously mentioned, the experiment was a success and it is much clearer now that nothing or no one can mess with the *Heracleum mantegazzianums*!

Bibliography:

- Allott, Andrew, and David Mindorff. *IB Biology*. Oxford University Press, 2014
- Brunning, Andy. "The Chemistry of Giant Hogweed and How It Causes Skin Burns."*Compound Interest*, 3 Aug. 2017, www.compoundchem.com/2017/08/03/gianthogweed/.
- Brunning, Andy. "The Chemistry of Stinging *Urtica dioica*s." *Compound Interest*, 4 June 2015, www.compoundchem.com/2015/06/04/*Urtica dioica*s/.
- Burg, Stanley P. "Ethylene in Plant Growth." *Proceedings of the National Academy of Sciences of the United States of America*, U.S. National Library of Medicine, Feb. 1973, www.ncbi.nlm.nih.gov/pmc/articles/PMC433312/.
- CBS News. "Giant Hogweed: 8 Facts You Must Know about the Toxic Plant." *CBS News*, CBS Interactive, 6 July 2011, www.cbsnews.com/pictures/giant-hogweed-8-facts-you-must-know-about-the-toxic-plant/3/.
- Herbpathy. "Giant Hogweed Herb Uses." *Giant Hogweed Herb Uses, Benefits, Cures, Side Effects, Nutrients*, Herbpathy, 2017, herbpathy.com/Uses-and-Benefits-of-Giant-Hogweed-Cid537.
- Lawley, Richard, et al. *The Food Safety Hazard Guidebook*. Royal Society of Chemistry, 2012.Chapter 2.1.2.3 Furocoumarins, page 239
- Math Is Fun. "Standard Deviation." *Standard Deviation Formulas*, Math Is Fun Advanced, 2014, www.mathsisfun.com/data/standard-deviation-formulas.html.
- Nearing, Brian. "State Program Aims to Destroy Plant That Can Scar, Blind." *TU*, Times Union, 10 May 2017, www.timesunion.com/tuplus-business/article/State-program-aims-to-destroy-plant-that-can-11137581.php.
- Perrone, Jane. "Giant Hogweed; Digging Deeper into the History of a 'Killer Weed'." *The Guardian*, Guardian News and Media, 15 July 2015, www.theguardian.com/lifeandstyle/gardening-blog/2015/jul/15/giant-hogweed-digging-deeper-into-the-history-of-a-killer-weed.
- Reid, Michael S. "Ornamentals English." *UC Postharvest Technology Center*, University of California, postharvest.ucdavis.edu/Commodity_Resources/Fact_Sheets/Datastores/Ornamentals_English/?uid=36&ds=801
- RHS. "Giant Hogweed." *RHS Gardening*, Royal Horticultural Society, www.rhs.org.uk/advice/profile?PID=458.

7. WHAT IS THE EFFECT OF PH CONCENTRATIONS ON THE EFFICIENCY OF IMMOBILIZED LACTASE

Author: Fiona Erskine
Moderated Mark: 23/24

Research Question

What is the effect of (4, 5, 6, 7, 8) pH concentrations on the efficiency of immobilised lactase by measuring the amount of glucose in the solution using a blood glucometer after 24 hours of exposure of lactase to lactose?

Gathered Information

Lactose intolerance is a chronic medical condition resulting from the inability of a certain percentage of the population to digest the lactose sugar, most commonly found in dairy products. Symptoms include bloating, cramping, and diarrhoea, which can lead to severe dehydration. "Approximately 65 percent of the human population has a reduced ability to digest lactose after infancy. Lactose intolerance in adulthood is most prevalent in people of East Asian descent, affecting more than 90 percent of adults in some of these communities." (Lactose Intolerance, 1). From this we can see that around 65% of the population suffers from some form lactose intolerance, the highest concentrations of which are found in Asia, Southern Africa, and South America. It is important to mention that these are also the regions with the highest incidence of poverty, unfortunately making lactose-free products an unaffordable luxury. By finding more efficient and exact methods of catabolising the lactose substrate, firms may be able to increase their productivity and sell products for cheaper prices without decreasing their revenue. This would make their products more accessible and affordable to people of lower socioeconomic status. I think this is a valuable lab to pursue because lactose intolerance is an incredibly uncomfortable and even dangerous ailment to deal with, especially if you don't have the resources to handle it properly. I think there should absolutely be more research on lactose intolerance and action taken to make products available to people in less than ideal situations.

Immobilised enzymes are enzyme units that have been adhesively attached to an insoluble material and cannot move freely about the substance, but are still able to carry out their function. Advantages to using immobilised enzymes include making the enzyme less sensitive to changes in temperature, giving the enzyme a better reaction rate, and increasing the stability of the enzyme. This is due to the binding of an enzyme to a surface, which allows the protein to become more stable and less likely to denature (Twadell, 1). In addition, immobilised enzymes can give greater yields of the product than free-moving enzymes, which is particularly interesting to the economics perspective of this lab (Markoglou, 215). Immobilised enzymes can also be easily separated from the product and even reused several times which saves both time and money for the producer of the enzyme-affected product (Zhang, 106). In this lab, the immobilised lactase enzyme will be used to break down lactose. There are many different methods of immobilising an enzyme, but here I will be using a process called crosslinking. This involves mixing sodium alginate and lactase enzyme solution with another solution of calcium chloride. The enzyme binds the insoluble $CaCl_2$ and forms beads. These beads can be taken from the solution and then introduced to the lactose sugar for metabolism (Amato, 1).

Lactase is used to break down lactose, a disaccharide sugar with the chemical formula $C_{12}H_{22}O_{11}$ into the universal monosaccharides of glucose $C_6H_{12}O_6$ and galactose

$C_6H_{12}O_6$ through a process called catabolism. Once broken down, the organism can use these monomers like building blocks to assemble the molecules necessary through a process called anabolism, or use them as resources for cellular respiration to obtain energy. Catabolism of lactose is done by the lactase enzyme. When lactose bonds to the active site of the enzyme, the enzyme's orientation allows it to split the molecule into glucose and galactose, which then separate from the lactase and enter the bloodstream (White, 1).Once in the bloodstream, several enzymes work to take the molecules through the different stages of cellular respiration: glycolysis, the citric acid cycle, and the electron transport chain. This aerobic process releases 36 ATP (Beck, What is). ATP is an organic chemical that supplies energy for cells to do work and maintain homeostasis. Most glucose that is broken down goes towards the making of ATP for cell energy. Lactose is a source of necessary sugar for many organisms, especially mammals in infancy. People who suffer from lactose intolerance may get their sugars from other foods as well, but not having the ability to eat any dairy products could seriously cut down on the glucose-intake of the organism. Research on lactose-free products, such as milk that has already been broken down into the monomers of glucose and galactose, allows lactose-intolerant patients maintain their necessary sugar levels.

All enzymes have a range of optimum pH and temperature, this is the pH or temperature where the enzyme works most effectively. Enzymes that are placed in temperatures lower than their optimum experience a decrease in their ability as the decrease in kinetic energy reduces the amount of collisions between the enzyme and the substrate. Without these collisions the enzyme cannot carry out its function as well, reducing the output. Temperatures that are higher than the enzyme's optimum can denature, or change the shape of the enzyme, making it ineffective. Lactase is efficient over a pretty wide range of temperatures but was found to function best at a temperature of 37°C (Hermida, 4836), around human body temperature. Optimum pH for lactase enzyme is known to be around 5.8 to 6.0 (Skovbjerg, 653), but the enzyme is still at least partially effective from a pH of about 4 to 8 (Mozumder, 10), many different trials have had different conclusions, there are even trials that have concluded that lactase works best around a pH of 5 (Gray, 729), (Ho, 1). When enzymes are exposed to environments that are either more acidic or more basic than their optimum, the 3D configuration of both the enzyme (lactase in this case) and the substrate (lactose) can change, making them unrecognisable to each other and unable to bond. This prevents the enzyme from carrying out its function (Sandhyarani, Effect of). In the human body, lactase is produced around the beginning of the small intestine. The organ has varying pHs as you move through it (Collins, Anatomy, Abdomen), but lactase is found at the beginning, right as the acidic contents from the stomach are passed over. The stomach has an incredibly low pH (2-4) so while the small intestine is more basic, this region must be able to function at lower pHs (Fallingbord, Intraluminal pH). The lactase enzyme is most commonly functioning in a pH of about 5.5 to 6, so it makes sense that this should be its optimum pH.

In an experiment to test at which pH lactase works best, several pH solutions will need to be prepared. To prepare a pH solution, buffer solutions must be used. A buffer solution is a mixture of a weak acid and its conjugate base (or vice versa). It is necessary when preparing pH solutions because of its ability to maintain its current pH, even when

121

a strong acid or strong base is introduced to the aqueous mixture. Buffer solutions equations for the pHs of 4, 5, 6, 7, and 8 can be seen further down.

Independent Variable		
pH (4, 5, 6, 7, 8)		
Dependent Variable		
The amount of glucose in the solution (mg/dL)		
Control Variables	**How to control**	**Why to control**
Temperature (°C)	Keep the solution at room temperature (20°C). Use thermometer. (± 0.1°C)	Increase or decrease in temperature can affect the efficiency of the enzyme. Use thermometer. (± 0.1°C)
Mass of enzyme	Measure out the same number of lactase balls for each experimental group with a digital balance (± 0.01g).	If there is more enzyme but the same amount of substrate, these enzymes will be more efficient at carrying out their function.
Concentration of substrate	Measure out the same amount of lactose for each experimental group using a digital balance (± 0.01g).	More substrate will allow for a greater concentration of the product in the final solution.
Time (minutes)	For each experimental group, take data on the amount of glucose present after 24 hours using a phone timer.	Data must be taken in the same time increments to measure the rate of efficiency of the enzyme.
Same technique for making enzyme balls	To make the balls, hold your elbow to the table to keep a constant height and push on the back of the syringe so that one drop falls per second	This will make all the balls the same size, ensuring that they all carry the same amount of the enzyme and all the solutions are exposed to the same amount of enzyme

Same stirring for alginate balls	Have the magnetic stirrer stir at a constant rate for the alginate balls	This is to ensure that all the balls end up the same size, and carry the same amount of enzyme
Same $CaCl_2$ solution for balls	Measure the solution concentration you want to use and use this amount for every batch of alginate balls you make	Since CaCl2 reacts with the alginate to produce the balls, different amounts of the compound would cause different size and concentration of enzyme of the balls
Same size beakers for lactose exposure	Use the same 250 ml flasks for each solution so that the lactase is exposed to equal amounts of lactose	By having the same size beakers and the same amount of each solution, this allows for equal amounts of each reactant to be exposed to each other
Same technique for checking glucose	Use the same blood glucometer and same brand of testing strips for every reading	This allows all the data to be taken with the same machine, so that if there are systematic errors they are widespread over the entire experiment and are a constant, allowing us to still have relative readings
Contamination of solutions	Use a permanent marker to mark each flask and each beaker with an indication of the contents inside	This prevents against human error and mixing two substances that aren't supposed to be mix and forcing a redo of the experiment
Volume control of solutions	Measure out the same amount of lactose solution and pH solution using a 50 ml graduated cylinder (± 1 ml)	This allows for each substance to be exposed to the same amount of reactant so that all the solutions start out equal to give conclusive results
Evaporation of solutions	Stopper each flask when you leave it in the fume cupboard overnight	This protects the solution from evaporation which could potentially evaporate one of the reactants causing unequal numbers of reactants which could alter one of the data sets

Mixture of all solutions	Place a magnetic bar inside all of the beakers and flasks and place equipment on a magnetic stirrer and adjust so that they spin at the same rate	This gives all of the reactants equal exposure to each other and allows a more equal distribution to promote better reaction and more opportunity for collisions

Hypothesis

If immobilised lactase enzyme is exposed to a pH that is either more basic or more acidic than a pH of 6, its efficiency at breaking down lactose will decrease because of the deviation from the optimum pH. This will cause the enzyme and substrate to change their shapes, decreasing their ability to bond and the enzyme's capability to carry out its function. The pH of 5 will be most efficient after pH 6 and the pH of 8 will be the least effective.

Method

A. Safety-

1. FLINN Safety Contract (FLINN Scientific Inc., 2017)
    a. "Dispose of all chemical waste properly. Never mix chemicals in sink drains. Sinks are to be used only for water and those solutions designated by the instructor. Solid chemicals, metals, matches, filter paper, and all other insoluble materials are to be disposed of in the proper waste containers, not in the sink. Check the label of all waste containers twice before adding your chemical waste to the container."
    b. "Labels and equipment instructions must be read carefully before use. Set up and use the prescribed apparatus as directed in the laboratory instructions or by your instructor."
    c. "Acids must be handled with extreme care. You will be shown the proper method for diluting strong acids. Always add acid to water, swirl or stir the solution and be careful of the heat produced, particularly with sulfuric acid."
    d. "Never return unused chemicals to their original containers."
    e. "Any time chemicals, heat, or glassware are used, students will wear laboratory goggles. There will be no exceptions to this rule!"
    f. Wear gloves when working with chemicals.
2. MSDS Chemicals (Material Safety Data Sheet, 2006)
    a. Sodium Alginate- No serious hazard with this chemical, avoid inhaling excessive amounts and wash off skin with soap and water.

b. Calcium Chloride- "May be harmful if swallowed. May cause severe respiratory and digestive tract irritation with possible burns. May cause severe eye and skin irritation with possible burns. May cause cardiac disturbances." As a precaution, make sure to "wear appropriate protective eyeglasses at all times" and "wash thoroughly after handling. Use with adequate ventilation. Do not get on skin or in eyes. Do not ingest or inhale." And always use cool water, heat evolved is significant. Avoid breathing dust, vapor, mist, or gas.

c. Potassium hydrogen phthalate- "May cause irritation to the eyes." Make sure to wear appropriate eye protection while handling.

d. Potassium dihydrogen phosphate- "May cause eye, skin, and respiratory tract irritation. The toxicological properties of this material have not been fully investigated." To avoid accidents "Wear appropriate gloves to prevent skin exposure. Wear appropriate protective eyeglasses or chemical safety goggles"

e. Sodium hydroxide- "Causes eye and skin burns. Causes digestive and respiratory tract burns. Hygroscopic (absorbs moisture from the air)." As a precaution, Wear chemical splash goggles and face shield. Wear butyl rubber gloves, apron, and/or clothing."

B. Materials List:

8-250ml beakers, 5-250ml flasks, 5 rubber stoppers, 1-200 ml graduated cylinder, 1- 50 graduated cylinder, 1-10 ml syringe, Goggles, Latex gloves, Digital balance, Weight boat, Metal spatula, Glass stirring rod, Bar magnet, Electromagnetic plate, Strainer, Permanent marker, 50 glucose testing strips, Electrochemical glucometer, 10 pH testing strips, Notebook, Sodium Alginate powder (2g), Calcium Chloride (1g), Lactase enzyme (20,000 enzyme units), Potassium hydrogen phthalate (25g), Potassium dihydrogen phosphate (25g), Sodium hydroxide ($1000cm^3$ of 0.1 molar), 250ml of lactase, 1 L distilled water

C. Procedure

1. Put on goggles and latex gloves
2. Preparing the immobilised enzymes
   a. Label a 250 ml beaker "solution A" with a permanent marker to avoid contamination.
   b. Zero out a digital balance with weight boat on it.
   c. Using a metal spatula, measure out 2g of sodium alginate powder
   d. Use a 1-250 ml graduated cylinder to measure 100 ml of distilled water
   e. Pour the alginate powder and distilled water into "solution A" beaker
   f. Mix with an electromagnetic stirrer and magnet until the powder is completely dissolved
      i. This takes about an hour

g. Now add the 2g of lactase enzyme to "solution A" and stir with electromagnetic stirrer and magnet again for 5 minutes

h. Zero out balance with weight boat on it.

i. Using a metal spatula, weight out 1g of calcium chloride

j. Label another 250 ml beaker "solution B" to avoid contamination

k. Mix calcium chloride with 100 ml distilled water using graduated cylinder to make "solution B"

l. Use a 1-10 ml syringe and take up 10 ml of solution A

m. Put a magnet into solution B and place on top of a electromagnetic stirrer

n. Add solution A to solution B drop by drop via the syringe

   i. To use syringe, hold your elbow to the table for stability and push on the top of the syringe slowly so that one drop comes out per second.

o. Wait for the beads to form (this should happen instantly)

   i. Repeat sub steps (l)-(o) twice

p. Filter the beads out of the solution using a strainer

q. Wash beads with distilled water

r. Store beads in a 250ml beaker with 100 ml DI water using graduated cylinder in a refrigerated environment

3. Preparing the pH solutions

   a. Use the equations below and put the following components into 5 different 250ml flasks (label the flasks 4, 5, 6, 7, and 8 to avoid contamination)

      i. Use a digital balance, weight boat, and metal spatula to measure the solids remember to zero out balance with weight boat on it each time

      ii. Use a graduated cylinder (1-50ml for NaOH and 1-250ml for H_2O) for the liquids

   b. Equations:

      i. pH of 4 - $C_8H_5KO_4$ (2.55g) + H_2O (250 ml)

      ii. pH of 5 - $C_8H_5KO_4$ (2.55g) + H_2O (193.5 ml) + NaOH (56.5cm^3 of 0.1 M)

      iii. pH of 6- KH_2PO_4 (1.70g) + NaOH (14cm^3 of 0.1 M) + H_2O (236 ml)

      iv. pH of 7- KH_2PO_4 (1.70g) + NaOH (72.75cm^3 of 0.1 M) + H_2O (177.25 ml)

      v. pH of 8- KH_2PO_4 (1.70g) + NaOH (116.75cm^3 of 0.1 M) + H_2O (133.25 ml)

   c. After all the components are in the flask, put the flasks on the electromagnetic stirrers with magnetics stir bar inside, let mix for 5 minutes

   d. Once the pH solution is made, test it with a pH testing strip to make sure it has the right concentration

   e. Place the rubber stopper inside the flask to avoid evaporation and contamination

      i. Make sure the pH is stored at room temperature (20-25°C)

4. Preparing the lactose solution

a. We want 0.5 molar concentration and 300ml ($0.3dm^3$) total of solution
b. Use concentration= # of moles/ volume to find the amount of lactose necessary
c. $0.5moles/dm^3$ = # of moles / $0.3dm^3$
d. # of moles = 0.15 moles x 342.3 (mass of 1 mole) = 51.35g for the solution
e. Using a 1-250ml graduated cylinder, measure out 300 ml of DI water
f. Using a digital balance, weight boat, zero out the balance with weight boat, then use and a spatula and measure out 51.35g of lactose
g. Mix the DI water and lactose into a 1-500 ml beaker and place on electromagnetic stirrer for 3 minutes to create a homogenous mixture

5. Take steps to avoid contamination
   a. Label 250 ml beakers pH 4, 5, 6, 7, and 8
   b. Assign a different plastic pipette each for each solution
   c. Remove used glucometer testing strips (50) once used
6. Pour the 50 ml of pH 4 buffer into the beaker using 1-50ml graduated cylinder
7. Put a weight boat on the digital balance and zero it
8. Use the balance to weigh out 10g of immobilised enzyme balls
9. Use a 1-50 ml graduated cylinder to measure 50 ml of the previously prepared lactase solution
10. Mix solution for 10 minute using an electromagnetic stirrer and magnet
    a. Repeat the same process using pHs 5, 6, 7, and 8
11. Place all the solutions in the fume cupboard for 24 hours
12. Taking data
    a. Use a pipette to take up the solution
    b. Wash your hands and put on a latex glove
    c. Allow one drop of the solution to fall onto your finger
    d. Dip the testing strip into the solution on your finger and plug it into the glucometer
       i. Repeat 10 times per pH solution
       ii. Make sure to use a different finger each time or change your gloves
    e. Record the amount of glucose detected in mg/dl in your notes.
       i. Repeat for each pH solution

Data Collection and Analysis

Table 2
pH Formula Table

Equations	
pH 4	$C_8H_5KO_4 + H_2O$
pH 5	$C_8H_5KO_4 + H_2O + NaOH$

pH 6	KH_2PO_4 + NaOH + H_2O
pH 7	KH_2PO_4 + NaOH + H_2O
pH8	KH_2PO_4 + NaOH + H_2O

Table 3
Qualitative Data Table

	Colour Change of Solution	Colour of the Balls	Consistency of the Solution	Structure of the balls
pH 4	no change	clear with a dark bubble inside	watery	balls stayed in tact
pH 5	no change	clear	watery	balls stayed in tact
pH 6	no change	clear	very sticky and viscous	balls stayed in tact
pH 7	no change	clear	watery	balls stayed in tact
pH8	changed darker	cloudy	jelly consistency	balls had exploded so the solution was mostly jelly

Table 4
Change in level of glucose/ mg/dL (±1 mg/dL) as a result of pH 4 solution after 24 hours in immobilised lactase balls.

Trials	Initial Glucose Concentration/mg/dL(± 1 mg/dL)	Final Glucose Concentration/mg/dL(± 1 mg/dL)	Final- Initial Glucose Concentration/mg/dL(± 1 mg/dL)
trial 1	0	10	10

trail 2	0	10	10
trial 3	0	lo	lo
trial 4	0	10	10
trial 5	0	11	11
trial 6	0	11	11
trial 7	0	10	10
trial 8	0	10	10
trial 9	0	10	10
trial 10	0	lo	lo
Mean			**10**
Median			10
Mode			10
STDEV			0.46
STE			0.15

* This is a representative table of raw data for all pH solutions. lo indicates a level of glucose that is less than 10 mg /dL. Because the glucometer used is meant for measuring the amount of glucose in blood, it was not equipped to measure very low levels of the substance. Because of this anomaly, these data points were excluded from the calculations. The mean, median, and mode show that there was central tendency since they are all close to the same number. The average reading was about 10 mg/dL and the STDEV and STE show that the results are consistent and there isn't much deviation from the mean.

Table 5

1 STDEV from the mean for change levels of glucose (mg/dL ±1 mg/dL) as a result of all pH solutions (4, 5, 6, 7, and 8) after lactose exposure to immobilised lactase balls for 24 hours.

pH					
	pH 4	pH 5	pH 6	pH 7	pH 8
mean + STDEV	11	14	21	12	0
mean -	10	13	20	1	0

STDEV					
range	10-11	13-14	20-21	1-12	0
% within 1 STDEV from mean	75%	90%	90%	30%	100%
mean + 2 STDEV				17	
mean - 2 STDEV				5	
range				5-17	
% within 2 STDEV from mean				30%	

*This is a representative table for 1 STDEV done for each pH solution. In the table above, we see that all of the results are above the 1 STDEV from the mean, except pH 7. This indicates that in all the other points, there is consistent and noteworthy data produced and that the results show central tendency. However, in pH 7, the model of glucometer used produced unclear results, especially lo values. Most of the values taken for this series were in the lo range, indicating that the amount of glucose detected was between 0 and 10. Theses data points had to be excluded from the calculations. Because of this, this series was left with only 3 valid data points, which is not enough to make a valid conclusion on the results.

Table 6
ANOVA table

Groups	Count	Sum	Average	Variance		
pH 4	8	82	10.25	0.21		
pH 5	10	133	13.3	0.46		
pH 6	10	205	20.5	0.50		
pH 7	5	31	6.2	32.20		
pH 8	8	0	0.0	0.00		
Source of Variation	SS	df	MS	F	P-value	F crit
Between Groups	2043.1	4	510.78	132.38	0.000	2.63
Within Groups	138.9	36	3.86			

Total	2182	40				

*Looking at the chart, we can see that the P-value is 100%, indicating that there is absolute confidence that there is a significant difference between each pH groups data.. The F value is much greater than the F critical value, which signifies that there is a significant difference between the data points. However, it is important to mention that for some groups, only 5 counts per group, which tells us that it is not as reliable.

Graph 1

Average Amount of Glucose mg/dL as a result of different pH solutions after 24 hours of lactose exposure to immobilized lactase.

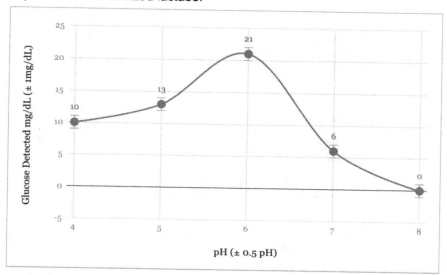

*As we can see, there is an exponential increase as the pH increases for 4 to 6, maxing at 21. This peak is followed by a rapid decrease in the amount of glucose produced, showing pH 7 with an average of 6.2 and pH with an average of 0. The error bars are based on the STE taken in tables 2-6. Because the width is quite narrow, this shows that there is a central tendency in the data points taken. There is obviously no overlap of data points for pHs 5-8; however, the pH 4 and 5 points seem close, the T-test below shows that the numbers are still significant.

Table 7

T-test between pH 4 and pH 5

	pH 5	pH 4
Mean	13.25	10.25
Variance	0.50	0.21
Observations	8	8
Pearson Correlation	0.22	

131

Hypothesized Mean Difference	0	
df	7	
t Stat	11.22	
P(T<=t) one-tail	0.000005	
t Critical one-tail	1.89	

In the T-test above, it is evident that the t-stat value is greater than the t-critical which indicates that there is a significant difference between the two data points collected. This means that although the error bars may seem to overlap on the graph, the two data points are statistically significant. The P-value one-tail shows that we have almost 100% confidence stating there is a significant difference.

Table 8

Table of Percentage Difference, Using pH 6 as Control

	pH 4	pH 5	pH 6	pH 7	pH 8
Average Value	10	13	21	6	0
Control Value	21	21	21	21	21
% Difference	52% decrease	38% decrease	0%	71% decrease	100% decrease

Using pH 6 as a control, we can see that all the other data points deviate from the control. pH 5 is the closest data point to the control and pH 8 was the furthest away. The order is as follows, pH 5, pH 4, pH 7, and pH 8 in terms of deviation from the control group.

Conclusion

In this experiment, I explored the effect of varying pHs on the efficiency of the lactase enzyme to catabolise lactose into glucose. In order to do this, I set up 5 solutions ranging from a pH of 4 to a pH of 8, immobilised the lactase enzyme in alginate balls, and allowed a constant amount of the lactose substrate to sit in the mixture overnight. The following day, I took readings on the glucose concentration in mg/dL present in the solution by using a blood glucometer.

In designing my experiment, I chose a pH of 6 as my control on which to base my conclusions. I chose the pH solution of 6 because research done before the experiment was conducted indicated that this was the pH that lactase was most efficient (Skovbjerg, 653). My experimental groups were the pH solutions of 4, 5, 7, and 8. I chose these because according to my research, the lactase enzyme is still effective at these pH, and unlikely to denature completely, although its ability to perform work is reduced. When looking at the graph of data gathered in the experiment, a parabolic curve is observed. This curve can be modelled by the equation $-2.93x^2 + 32.4x - 73.4$. Since the coefficient of the first term is negative, this indicates that the parabola is opening downwards in relation to the y axis and has a maximum value, or peak. This point is representative of the pH where the lactase enzyme was most efficient; this graph showed that optimum was reached between pH of 5.5 and 6. This is further supported through the examination of the averages in the amount of glucose present in each solution. The concentration of glucose in the pH of 6 solution was about 21 mg/dL. This average is 8 mg/dL greater than the next highest glucose concentration average in the experiment, which was achieved by the pH of 5 solution. pH 5 showed a 38% difference from the control value, followed by a pH of 4 which had 52% difference from the control value and an average of 10 mg/dL. pH 7 had a 71% difference and a mean of about 6 md/dL, and pH 8 had 100%, in fact, no glucose was detected in this solution at all with an average of 0 mg/dL.

This data indicates that the efficiency of the lactase enzyme is reduced the more acidic the pH becomes, but is still active. This is in agreement with the current science that supports that the lactase enzyme is present in small intestine in humans. The small intestine has pHs that become more alkaline as waste moves away from the stomach. Where the stomach can have pHs between 2 and 4, the small intestine has a pH of about 6 (Fallingbord, Intraluminal pH). It makes sense that the lactase would work best at pH 6 but still be effective at pHs of anywhere from 4 to 5 so that the enzyme would be able to digest lactose as it moves away from the highly acidic stomach region, which is seen in the experiment in question. On the other hand, a sharp decrease in the ability of lactase in environments with pH higher, or more basic, is seen through the results of pH 7 and 8. While the blatantly noticeable decrease in the pH 7 data may be due to limitations in the blood glucometer, it may be also be attributed, in part, to the fact that lactase is found closer to the beginning of the small intestine, where all the acidic products have just exited the stomach. While this organ does reach pHs of 7 towards its other end, the small intestine can grow up to 15 feet long (Collins, Anatomy, Abdomen) and lactase is unlikely to be exposed to the pH (Evans, Measurement of). Because of this, it is unlikely that the enzyme will work efficiently in this pH, due to the fact that it doesn't need to in the body. Finally, because the glucometer detected no glucose present in the pH of 8 solution, we can reasonably assume that a significant percentage of the enzyme was completely denatured, meaning that the 3D configuration of the enzyme changed beyond recognition for the substrate, compromising its ability to break down the lactose. This could be due to the lactase enzyme never being exposed to a pH of 8 in the small intestine and being too far away from its optimum pH.

In terms of confidence in the data collected, the *p-value* provided by the ANOVA table allows me 100% confidence that the data taken has a significant difference between each pH solution. Along with the *p-value*, the *f-value* -132.28 is obviously greater than the *f-critical* value 2.63. This supports that there is a significant difference in the data taken. However, because pH 7 only had a count of 5 data points, this makes the results less reliable and inconclusive for the pH 7 data. In addition, when looking at the graph of the data points, it looks like the error bars of the pH of 4 and pH of 5 might overlap. In order to confirm that they don't, I conducted a one-tailed t-test between the averages of the glucose concentration of the solutions. The t-test shows that the *t-stat* value 11.22 is greater than the *t-critical* value 1.89. The *p-one tail* also allows for 100% confidence that the pH 4 and 5 data points are statistically significantly different means.

Based on the confidence levels in the data collected, there is a strong argument that the experiment shows a parabolic curve with an optimum at pH 6 and that all data points taken were statistically significant except inconclusive data on pH 7. In addition, both currently accepted science and logic in considering where the lactase enzyme is found in the human body support the conclusion that pH 6 is the optimum pH for the lactase enzyme. It also supports that the enzyme functions at more acidic pHs, though at a reduced rate, and can completely denature when exposed to environments of more alkaline pH. My original hypothesis stated that derivation from the optimum pH of 6 will lead to reduced capacity for the enzyme, my data does support this hypothesis but also allows for more information regarding the degree to which derivation from mean can affect the ability of the enzyme to do work and how its function is inhibited depending on if you increase or decrease the pH.

Evaluation

Table 9

Strengths	Control Variables	In my experiment, both temperature and the duration that the lactose was exposed to lactase in the solution were controlled well. I did this my taking the temperature of the solutions before and after the 24 hour period and storing them in the same fume cupboard. All solutions maintained the same temperature. In addition, all the solutions were measured over a 24 hour period. All solutions had 50 ml of pH solution, 10g of lactase enzyme balls, and 50 ml of 0.5 molar lactose solution. A glucometer was used to detect glucose levels in each of the solutions and every solution was mixed using a magnetic stirrer plate which rotated at the same speed.
	Central Tendency	Evidence - Mean, Median, and Mode were very close in each group except for the pH of 7 solution. 1 STD from the Mean- all groups were 1 STDEV from

134

		the mean except for the group with the pH of 7. Width of Error Bars- the error bars showed a width of 1 mg/dL, indicating small STE and precise data collection
Weaknesses and Improvements	Random	**Instrument Resolution**- to measure the concentration of glucose present in the solution after 24 hours, I used a blood glucometer, which is designed to measure the concentration of glucose in human blood for diabetics. Because of this intended purpose, it was not equipped to measure very low levels of glucose. Any measurement taken between 0-10 mg/dL produced "*lo*" on the screen. This data was unusable in my results because there is no way to know where in that range the reading fell. Most of the data for the pH solution of 7 fell in this range (most likely between 8-10), so most of that data was unusable. In addition, when the glucometer detected no glucose, "Error 11" was shown on the screen. Online information told us that this indicated 0 mg/dL detected but trace amounts could have been present in the solution without us knowing. The glucometer had an uncertainty of 1 mg/dL, which is higher than I would've like as it gave me a relatively high percentage error as I was working with lower numbers like 10,13, and 21. *Improvement*- To improve this experiment, I would definitely invest in a more sophisticated glucometer, especially one that would measure very small amounts of glucose (between 0 and 10 mg/dL). This would allow for more data points and a greater general accuracy
	Random	**Incomplete Definition**- When immobilising the lactase enzyme inside the alginate balls, I held my elbow to the table for balance and a constant height. However, because I did this for hours, I'm sure my position shifted a couple times and I definitely observed the balls at different sizes, presumably with different amounts of lactase enzyme. *Improvement*- To improve this, I would set up a ring stand so that all the alginate would fall from the same height at the same rate. This would also not require as much physical exertion from the person doing the experiment

	Systematic	**Instrument Drift**- Because I prepared my own pH solution, there is some uncertainty on the actual pH that was used. I tried to use a vernier pH probe, but it kept predicting my pH 2 levels higher than the one that was predicted by the pH strips, due to this drift, I switched to only using pH strips to measure the pH of my solution, which had a pretty large uncertainty of 0.5. *Improvement*- In order to improve this aspect, I would probably calibrate the pH probe (or buy a new one) and use that to measure the pH of the solutions. The probe has a much smaller uncertainty and probably would've given more accurate results.
	Limitations	**Concentrations of Solutions**- When preparing the pH solutions, I had to use the potassium hydrogen phthalate, potassium dihydrogen phosphate, and sodium hydroxide (0.1 molar) that was already in the lab. While I trust that these were precise, it is possible that the concentrations of these solutions were exposed to some human error. In addition, all the containers had been previously used and exposed to oxygen and possibly even some water in the case of potassium hydrogen phthalate. The sodium hydroxide also had some unidentified substance growing in the center of it, these factors could have affected the solutions.
		Equipment- As is previously mentioned, I ran into difficulty with the glucometer, the pH probe, and pH strips when conducting the experiments in terms of the accuracy and the uncertainty of the equipment.
Lab Extensions		If I were to choose a lab extension for this topic, I would take the data that I gathered in this experiment where I found that a pH of 5.5-6 was optimum and use this to make a more precise experiment. Instead of measuring the effect pH has on the enzyme between 4 and 8, I would narrow down my range to 5.5-6.5, using pH 6 as my control and experimental groups of pH 5.5, 5.75, 6.25, 6.5. While it would be difficult to measure these pHs, if we got a sophisticated pH probe, I think this would be feasible. Doing this experiment would help me to further narrow down the optimum pH for the enzyme and provide better indication of the pH needed to most efficiently break down lactose for lactose intolerant

Bibliography

Amato, Andrea. "Immobilization of Lactase Enzyme on Alginate Beads - A Quick Test." *Chemical Education Xchange*, 8 Jan. 2019,
www.chemedx.org/article/immobilization-lactase-enzyme-alginate-beads-quick-test.

Beck, Kevin. "What Is the Role of Glucose in Cellular Respiration?" Sciencing, 5 Apr. 2019,sciencing.com/role-glucose-cellular-respiration-6507636.html.

Collins, Jason T. "Anatomy, Abdomen and Pelvis, Small Intestine." StatPearls [Internet]., U.S. National Library of Medicine, 5 Apr. 2019,
www.ncbi.nlm.nih.gov/books/NBK459366/.

Evans, D F, et al. "Measurement of Gastrointestinal PH Profiles in Normal Ambulant Human Subjects." Gut, U.S. National Library of Medicine, Aug. 1988,
www.ncbi.nlm.nih.gov/pmc/articles/PMC1433896/.

Fallingborg, J. "Intraluminal PH of the Human Gastrointestinal Tract." Danish Medical Bulletin, U.S. National Library of Medicine, June 1999,
www.ncbi.nlm.nih.gov/pubmed/10421978.

Gray, Gary M. "Intestinal β-Galactosidases II. Biochemical Alteration in Human Lactase Deficiency." *National Center for Biotechnology Information,*The American Society for Clinical Investigation , Apr. 1969 www.ncbi.nlm.nih.gov/pmc/articles/PMC322277/.

Hermida, Carmen, et al. "Optimizing the Enzymatic Synthesis of β-d-Galactopyranosyl-d-Xyloses for Their Use in the Evaluation of Lactase Activity in Vivo." *Bioorganic & Medicinal Chemistry*, vol. 15, no. 14, 15 July 2007, pp. 4836–4840., doi: 10.1016/j.bmc.2007.04.067.

Ho, John N. "The Effect of PH on Lactase." *California State Science Fair*, California Science and Engineering Fair, 2009,
cssf.usc.edu/History/2009/Projects/S0410.pdf.

Markoglou, Nektaria, and Irving W Wainer. "Immobilized Enzyme Reactors in Liquid Chromatography: On-Line Bioreactors for Use in Synthesis and Drug Discovery." *ScienceDirect*, Elsevier Science B.V., 2 Sept. 2007,
www.sciencedirect.com/science/article/pii/S1567719203800086.

Mozumder, N. H. M. R., et al. "Study on Isolation and Partial Purification of Lactase(β-Galactosidase) Enzyme from Lactobacillus Bacteria Isolated from Yogurt."

Journal of Scientific Research, vol. 4, no. 1, 2011, p. 239.,
doi:10.3329/jsr.v4i1.8478.NIH. "Lactose Intolerance - Genetics Home Reference - NIH." *U.S. National Library of*

Medicine, National Institutes of Health, 5 Mar. 2019,
ghr.nlm.nih.gov/condition/lactose-intolerance.Precision Laboratories. "Oops! Errors in PH Measurement." Precision Laboratories, 31 Oct.2016,
preclaboratories.com/oops-errors-in-ph-measurement/.

Sandhyarani, Ningthoujam. "Effect of PH on Enzymes." *BiologyWise*, BiologyWise, 22 Feb. 2018, biologywise.com/ph-effect-on-enzymes.

Skovbjerg, Hanne, et al. "Purification and Characterisation of Amphiphilic Lactase/Phlorizin Hydrolase from Human Small Intestine." *FEBS Letters*, Wiley-Blackwell, 3 Mar. 2005,
febs.onlinelibrary.wiley.com/doi/full/10.1111/j.1432-1033. 1981.tb05193.x.

CPSIA information can be obtained
at www.ICGtesting.com
Printed in the USA
BVHW011253120421
604745BV00017B/120